Geometry Essentials

Geometry Essentials

Amanda Burke

STATES
ACADEMIC PRESS
www.statesacademicpress.com

Geometry Essentials
Amanda Burke
ISBN: 978-1-63989-238-9 (Hardback)

STATES
ACADEMIC PRESS

Published by States Academic Press,
109 South 5th Street,
Brooklyn, NY 11249, USA

Cataloging-in-Publication Data

Geometry essentials / Amanda Burke.
p. cm.
Includes bibliographical references and index.
ISBN 978-1-63989-238-9
1. Geometry. 2. Mathematics. I. Burke, Amanda.
QA445 .G46 2022
516--dc23

For more information regarding States Academic Press and its products, please visit the publisher's website www.statesacademicpress.com

Table of Contents

Preface

The branch of mathematics which deals with the properties of space is known as geometry. It studies the dimensions, shape and position of figures in either a 2-dimensional or 3-dimensional space. Some basic concepts of this field include line, surface, angle, plane, curve and similarity. The subject of geometry is further divided into various sub-disciplines such as algebraic geometry, differential geometry, discrete geometry, convex geometry, Euclidean geometry and topology. Geometry finds its applications across various fields such as physics, art, architecture, etc. The topics included in this book on geometry are of utmost significance and bound to provide incredible insights to readers. Different approaches, evaluations and methodologies on geometry have been included herein. Coherent flow of topics, student-friendly language and extensive use of examples make this book an invaluable source of knowledge.

A detailed account of the significant topics covered in this book is provided below:

Chapter 1- Geometry is a branch of mathematics dealing with the study of sizes, angles and dimensions of shapes like circles, triangles, squares, etc. Differential geometry, Euclidean geometry, Non-Euclidean geometry, algebraic geometry, complex geometry, computational geometry, analytic geometry, etc. are the different types of geometry which are reviewed in this chapter.

Chapter 2- Geometry has its own fundamental concepts which are crucial for gaining a better understanding of the subject. Axiom, point, line, line segments, plane, ray, angle, curve, manifold, length, area and volume are the fundamental geometrical concepts examined in detail in this chapter.

Chapter 3- A polygon may be defined as a flat, two-dimensional closed shape with straight sides. Polygons come in all shapes and number of sides. Regular polygons, irregular polygons, convex polygons, concave polygons, quadrilateral polygons, pentagon polygons, etc. are a few examples. This chapter has been carefully written to provide the reader with a clearer understanding of polygons and their different types.

Chapter 4- A triangle is a two-dimensional, three-sided closed polygon. Triangles can be sorted into two groups based on their sides and angles. Equilateral triangles, isosceles triangles and scalene triangles are based on their sides. Triangles based on their angles are right triangles, obtuse triangles and acute triangles. This chapter provides a complete overview of triangles for the reader's benefit.

Chapter 5- A circle is a closed, two-dimensional round shaped figure that has no corner or edges. While it may certainly be one of the easiest shapes to reproduce freehand, a better understanding of circles can be achieved only after mastering its basics. The different parts of a circle like the radius, chord or diameter, its perimeter, area and its properties are the topics addressed in this chapter.

Chapter 6- A quadrilateral is a closed, two-dimensional shape with four straight sides. A quadrilateral with four equal sides is called a square. A rectangle is a quadrilateral with four right angles and parallel opposite sides of equal lengths. A quadrilateral with at least one pair of parallel sides is a trapezoid. This chapter has been carefully written to provide a comprehensive understanding of the subject matter.

Chapter 7- This chapter's focus is on three-dimensional, solid figures in geometry. These figures are all solid, three-dimensional shapes with multiple polygonal faces. In the case of polyhedrons distinctive flat and curved surfaces, pointed at the top as in cones or round solids are defined in three-dimensional space like spheres. This chapter aims to provide a better understanding by closely examining these topics.

It gives me an immense pleasure to thank our entire team for their efforts. Finally in the end, I would like to thank my family and colleagues who have been a great source of inspiration and support.

Amanda Burke

1

Geometry: An Introduction

Geometry is a branch of mathematics dealing with the study of sizes, angles and dimensions of shapes like circles, triangles, squares, etc. Differential geometry, Euclidean geometry, Non-Euclidean geometry, algebraic geometry, complex geometry, computational geometry, analytic geometry, etc. are the different types of geometry which are reviewed in this chapter.

Geometry is the branch of mathematics that deals with shapes, angles, dimensions and sizes of a variety of things we see in everyday life. In Euclidean geometry, there are two-dimensional shapes and three-dimensional shapes.

In plane geometry, 2 shapes such as triangles, squares, rectangles, circles are also called flat shapes. In solid geometry, 3d shapes such as a cube, cuboid, cone, etc. are also called solids. The basic geometry is based on points, line and plane.

Differential Geometry

A branch of geometry dealing with geometrical forms, mainly with curves and surfaces, by methods of mathematical analysis. In differential geometry the properties of curves and surfaces are usually studied on a small scale, i.e. the study concerns properties of sufficiently small pieces of them. Properties of families of curves and surfaces are also studied.

Differential geometry arose and developed in close connection with mathematical analysis, the latter having grown, to a considerable extent, out of problems in geometry. Many geometrical concepts were defined prior to their analogues in analysis. For instance, the concept of a tangent is older than that of a derivative, and the concepts of area and volume are older than that of the integral.

The Theory of Curves

The major subject of the theory of curves are the so-called differentiable curves. These are curves which can locally be specified by equations of the type,

$$x = x(t),\ y = y(t),\ z = z(t),$$

where $x(t), y(t), z(t)$ are sufficiently regular functions of a parameter t. The degree of differentiability of the curve is given by the degree of differentiability of $x(t)$, $y(t)$ and $z(t)$. For one and the same curve there are uncountably many ways of parametric description of the type $x = x(t),\ y = y(t),\ z = z(t)$. Of these, the so-called natural parameterization, in which the length of an arc of the curve, counted from some given point, serves as the parameter, is especially important. A point on a curve is said to be regular if, by a suitable choice of a Cartesian coordinate system x, y, z, the curve can be defined, in a neighbourhood of that point, by equations of the form,

$$y = y(x), z = z(x),$$

where y(x) and z(x) are differentiable functions. Otherwise it is called singular. In differential geometry the study of a curve mainly concerns a neighbourhood of a regular point. For a point on a curve defined by the general equation $x = x(t),\ y = y(t),\ z = z(t)$ to be regular, it is necessary and sufficient that the inequality,

$$x'2 + y'2 + z'2 \neq 0$$

holds at this point.

A number of fundamental concepts in the theory of curves are based on the concept of contact of sets, which may be explained as follows. Let M and m be two sets with a common point o. One says that the sets M and m have a contact of order α ≥ 1 at o if,

$$\frac{\delta(X)}{|Xo|^{\alpha}} \to 0$$

as X → o, where δ(X) is the distance of a point X of the set M from m. If M is a curve while m is a straight line passing through a point o of the curve, then if α ≥ 1, the contact condition defines m to be the tangent to the curve at o.

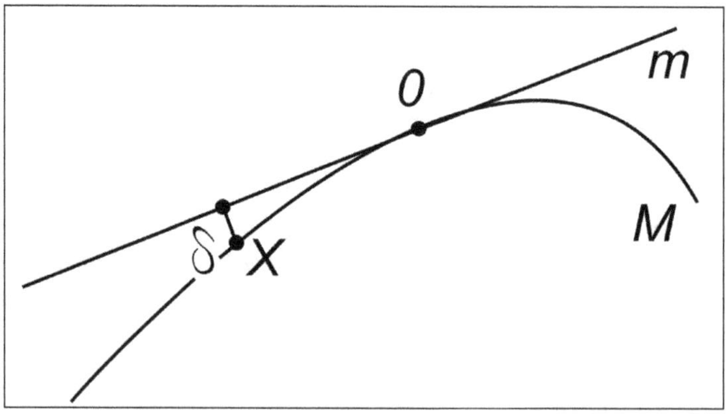

A smooth (differentiable) curve has a definite tangent at each of its points. The direction of the tangent at a point to of a curve specified by equation $x = x(t),\ y = y(t),\ z = z(t)$ coincides with the direction of the vector $[x'(t_o), y'(t_o), z'(t_o)]$. In differential geometry the equations of the tangent are derived for the various ways in which the curve is analytically specified. In particular, if the curve is defined by equations $x = x(t),\ y = y(t),\ z = z(t)$, the equations of the tangent at the point corresponding to the value to of the parameter are,

$$(X, Y, Z) = (x_o + ux'_o,\ y_o + uy'_o,\ z_o + uz'_o),\ u \in R,$$

where the subscript "o" indicates the values of the functions x, y and z and their derivatives at the point t_0. If m is a plane passing through a point o of the curve M, the contact condition for $\alpha \geq 2$ defines the osculating plane to the curve.

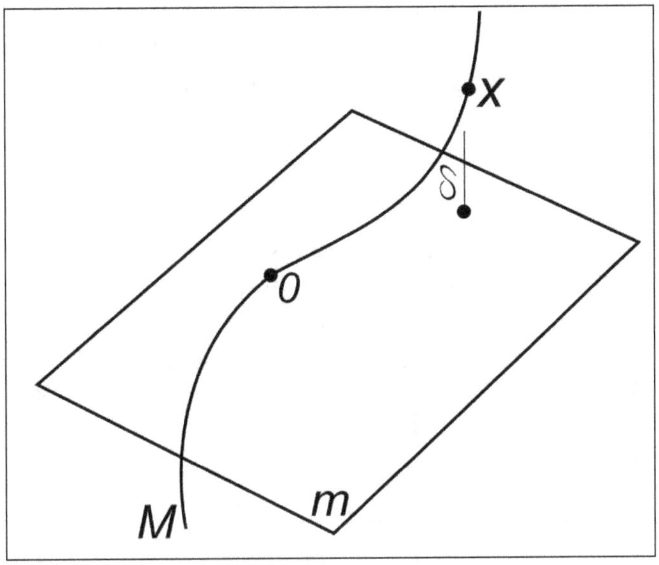

A twice differentiable curve has an osculating plane at each of its points. It is either unique or else any plane passing through the tangent to the curve is osculating.

When moving along a curve, the tangent to the curve undergoes a rotation. The velocity of this rotation, as the motion along the curve takes place at velocity one, is called the curvature of the curve. If the curve has been parameterized as in equation $x = x(t),\ y = y(t),\ z = z(t)$, its curvature is given by the formula,

$$k_1 = \frac{|\mathbf{r}'(t) \times \mathbf{r}''(t)|}{|\mathbf{r}'(t)|^{3/2}},$$

where $\mathbf{r}(t)$ is the vector function with coordinates x(t), y(t), z(t). Straight lines, and only straight lines, have a curvature which is everywhere equal to zero. A twice differentiable curve has a unique osculating plane at all points at which the curvature is non-zero. As one move along the curve in a neighbourhood of such a point, the osculating plane rotates, and the tangent to the curve is the instantaneous axis of this rotation. The velocity of rotation of the osculating plane, for a uniform motion along the curve at velocity one, is called the torsion of the curve. The sign of the torsion depends on the direction of the rotation. A thrice differentiable curve has a definite torsion at each of its points where the curvature is non-zero. If the curve has been parameterized as in equation $x = x(t),\ y = y(t),\ z = z(t)$, its torsion is given by the formula,

$$k_2 = \frac{(\mathbf{r}', \mathbf{r}'', \mathbf{r}''')}{|\mathbf{r}' \times \mathbf{r}''|^2}.$$

The torsion of a planar curve is zero at all points. Conversely, a curve with torsion identically equal to zero is planar. A straight line perpendicular to the tangent and passing through the point of contact is known as a normal to the curve. The normal lying in the osculating plane is called the principal normal, while the normal which is perpendicular to the osculating plane is known as the binormal. The figure constructed out of the tangent, the principal normal and the binormal, and also given by the three planes containing pairwise these straight lines, is called the Frénet frame (natural frame; also called Frénet trihedron and natural trihedron). If the edges of this frame at the given point are considered as the axes of a Cartesian coordinate system, then the equation of the curve in a natural parameterization in a neighbourhood of this point will have the form,

$$x = s + \ldots, y = \frac{k_1}{2}s^2 + \ldots, z = -\frac{k_1 k_2}{6}s^3 + \ldots,$$

where k_1 and k_2 are the curvature and the torsion of the curve at this point. Figure below represents the projections of the curve on the faces of the natural frame near a point with a non-zero curvature and a non-zero torsion.

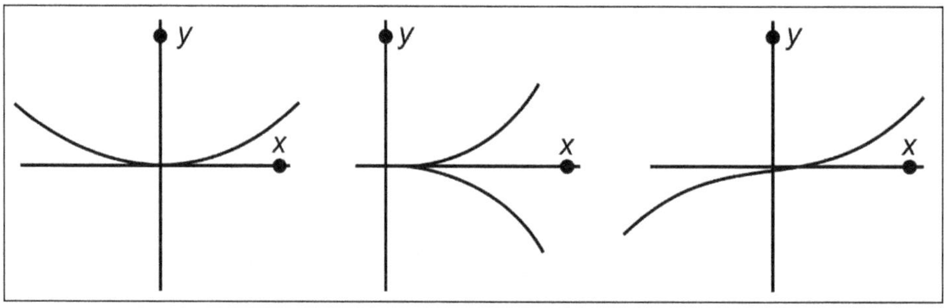

The unit vectors τ, ν, β of the tangent, of the principal normal and of the binormal to a curve vary during a motion along the curve. A suitable choice of directions of these vectors, in accordance with the definitions of the curvature and the torsion, yields the formulas,

$$\tau' = k_1 \nu, \beta' = k_2 \nu, \nu' = -k_1 \tau - k_2 \beta,$$

where the prime indicates differentiation with respect to the arc of the curve. The formulas above are known as the Frénet formulas. A curve with non-zero curvature is defined, up to its location in space, by specifying its curvature and its torsion as functions of the arc length s of the curve. This is why the system of equations,

$$k_1 = k_1(s), k_2 = k_2(s)$$

is also known as the system of natural equations of the curve.

Planar curves, i.e. curves lying in a plane, are an important class of curves. In the case of planar curves it is possible to distinguish between the directions of rotation of the tangent moving along the curve, so that the curvature can be given a sign depending on the direction of this rotation. The curvature of a planar curve defined by the equations $x = x(t), y = y(t)$ is given by the formula,

$$k = \pm \frac{y''x' - x''y'}{(x'2 + y'2)^{3/2}}.$$

The plus or minus sign is chosen in accordance with a convention, but must be the same all along the curve. The important concept of the osculating circle is introduced for planar curves. This is a circle with a contact of order $\alpha \geq 2$ with the curve.

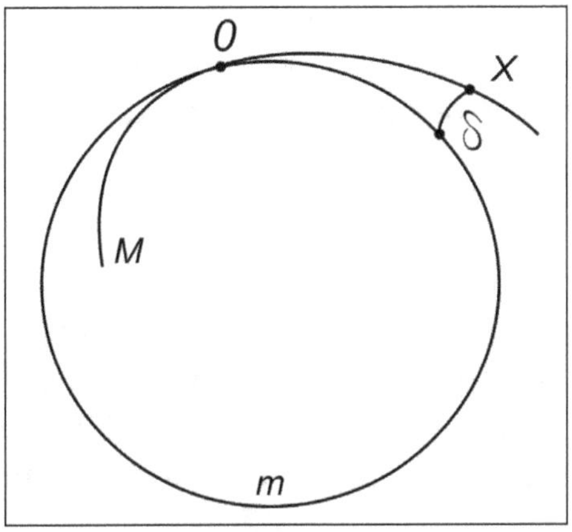

It exists at each point of a twice differentiable curve with a non-zero curvature. The centre of the osculating circle is known as the centre of curvature, while its radius is known as the radius of curvature. The radius of curvature is the quantity inverse to the curvature. The locus of the centres of curvatures of a curve is known as the evolute. The curve which orthogonally intersects the tangents to the curve is called the evolvent. The evolute of the evolvent of a curve is the curve itself.

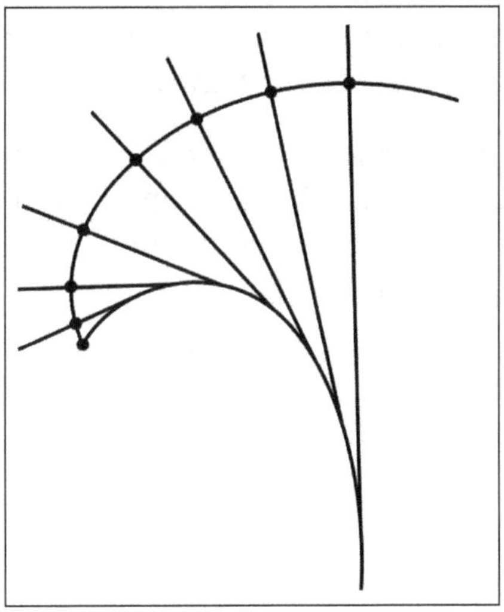

A curve Υ is called the envelope of a family of curves Υ_α, depending on a parameter α, if at each of its points it is tangent to at least one curve of the family, and if each of its segments is tangent to an infinite set of these curves.

The Theory of Surfaces and its Generalizations

The theory of surfaces usually deals with differentiable surfaces. These are surfaces which can be locally defined by equations of the type,

$$x = x(u,v), y = y(u,v), z = z(u,v),$$

where $x(u, v)$, $y(u, v)$, $z(u, v)$ are differentiable functions of the parameters u and v. The degree of differentiability of the surface is given by the corresponding degree of differentiability of these functions. A surface has uncountable many ways in which it can be defined by parametric equations such as equation above. A point of a surface is said to be regular (ordinary) if, in a neighborhood of it, a suitable choice of the coordinates x, y, z enables one to give the surface in the form,

$$z = z(x, y),$$

where z(x, y) is a smooth function; otherwise, the point of the surface is said to be singular. In differential geometry the study of the surface is mainly conducted in a neighbourhood of the regular (ordinary) points. For a point (u_0, v_0) of a surface given by the equations $x = x(u,v), y = y(u,v), z = z(u,v)$, to be regular, it is necessary and sufficient that the rank of the matrix,

$$\begin{pmatrix} x_u & y_u & z_u \\ x_v & y_v & z_v \end{pmatrix}$$

at this point be equal to two. If a surface is given by equations of the type $x = x(u,v), y = y(u,v), z = z(u,v)$, it is usually assumed, without stating this explicitly, that this condition is satisfied. If either u or v is fixed, the equations $x = x(u,v), y = y(u,v), z = z(u,v)$, define curves on the surface. Such curves are said to be coordinate lines on the surface. The parameters u and v are called surface coordinates or curvilinear Gaussian coordinates.

The concept of a tangent plane to a surface is defined in terms of the concept of contact. It is a plane passing through a point on the surface having contact of order $\alpha \geq 1$ with the surface at that point. A smooth (differentiable) surface has a unique tangent plane at each (regular) point. The tangent plane of a surface defined by equations $x = x(u,v), y = y(u,v), z = z(u,v)$, under condition $z = z(x,y)$, at the point (u_0, v_0), is defined by the equation,

$$\begin{vmatrix} X - x^o & Y - y^o & Z - z^o \\ x_u^o & y_u^o & z_u^o \\ x_v^o & y_v^o & z_v^o \end{vmatrix} = 0,$$

where the superscript o indicates that the functions x(u, v), y(u, v), z(u, v) and their derivatives are evaluated at the point (u_0, v_0). The straight line which passes through a point of the surface and is perpendicular to the tangent plane at this point is called the normal to the surface. If **r**(u, v) is a vector with coordinates x(u, v), y(u, v), z(u, v), then,

$$\mathbf{r}_u \times \mathbf{r}_v$$

has the direction of the normal to the surface.

The important concept of an osculating paraboloid is introduced for surfaces. This is a paraboloid the axis of which is the normal to the surface at the given point and with contact of order $\alpha \geq 2$ to the surface at this point. A twice differentiable surface has a unique osculating paraboloid at each of its points, which may degenerate to a parabolic cylinder or a plane. If the surface is described in Cartesian coordinates, with the given point as the coordinate origin, while the tangent plane at this point is taken as the xy- plane, the equation of the surface in a neighbourhood of this point will be,

$$z = f(x,y),$$

while the equation of the osculating paraboloid at this point is,

$$z = \frac{1}{2}\left(f_{xx}x^2 + 2f_{xy}xy + f_{yy}y^2\right)$$

(the derivatives of the function f being taken at the point of contact). Depending on the type of the osculating paraboloid, the points of the surface are divided into elliptic points, hyperbolic points, parabolic points, and flat points. The importance of the osculating paraboloid is due to the fact that it reproduces the form of the surface up to infinitesimals of the second order (the tangent plane reproduces this form up to infinitesimals of the first order).

The osculating paraboloid is used to introduce the concept of conjugate directions on a surface. Namely, two directions on a surface at a given point are said to be conjugate if the straight lines generated by these directions are conjugate with respect to the osculating paraboloid at this point. Orthogonal conjugate directions are called principal. At a given point on the surface there are, as a rule, two principal directions. Flat points and special elliptic points, at which all directions are principal, form an exception to this rule. A curve on a surface for which the tangent direction is principal at all points is said to be a curvature line. At non-elliptic points of the surface there exist self-conjugate directions. They are known as asymptotic directions. A curve on a surface for which the direction is asymptotic at all points is known as an asymptotic line.

The concept of the envelope of a family of surfaces is introduced similarly to that of the envelope of a family of curves in a plane. However, the family of surfaces may be a one-parameter or a two-parameter family. In the theory of surfaces, the envelope of a one-parameter family of planes is of special significance.

An important role in the theory of surfaces is played by two differential quadratic forms, the fundamental forms of a surface. Let **r**(u, v) denote the vector of a point on a surface, and let **n**(u, v) denote the unit vector of the normal to the surface; the fundamental forms are then written as,

$$I = d\mathbf{r}^2,\ II = -d\mathbf{r}d\mathbf{n}.$$

The coefficients of the first and the second fundamental form are usually denoted by E, F, G and L, M, N, respectively. The first fundamental form yields the distance on the surface between a point (u, v) and an infinitesimally close point $(u+du, v+dv)$:

$$ds = \sqrt{Edu^2 + 2Fdudv + Gdv^2}.$$

The length of a curve defined on the surface by the equations u = u(t), v = v(t) is computed with the aid of the first fundamental form:

$$L = \int_{t_0}^{t} \sqrt{Eu'2 + 2Fu'v' + Gv'2dt}.$$

The first fundamental form of the surface defines the angles between curves on the surface. In particular: the formula,

$$\cos\theta = \frac{F}{\sqrt{EG}}$$

is valid for the angle θ between the coordinate lines u = const, v = const at a point where they intersect. It is seen, accordingly, that the coordinate net on the surface is orthogonal if F = 0. The area of a piece of the surface is also given by the first fundamental form and, for a domain Ω on the surface, is computed by the formula,

$$S = \iint_{\Omega} \sqrt{EG - F^2}\, du dv.$$

The second fundamental form characterizes the twist of the surface in space. In fact, the ratio of the second to the first fundamental form,

$$k = \frac{L du^2 + 2M du dv + N dv^2}{E du^2 + 2F du dv + G dv^2},$$

represents the curvature of the planar section normal to the surface, drawn in the direction du : dv. There is a simple connection between the curvature of a curve lying on the surface and the curvature of the normal section of the surface through the tangent of the curve (the Meusnier theorem). The extremal values of the normal curvature of a surface at a given point are known as the principal and are reached in the principal directions. The normal curvature of a surface in an arbitrary direction can be expressed in terms of the principal curvatures and the angles formed by this direction with the principal directions (the Euler formula). The principal curvatures k_1 and k_2 are defined by the equation,

$$\begin{vmatrix} Ek - L & Fk - M \\ Fk - M & Gk - N \end{vmatrix} = 0.$$

Their mean value is known as the mean curvature of the surface. An important class of surfaces is constituted by surfaces of mean curvature zero — the so-called minimal surfaces. They are distinguished by the fact that a sufficiently small piece of such a surface has least area among all surfaces with the same boundary. The product $K = k_1 k_2$ of the principal curvatures is known as the Gaussian curvature of the surface:

$$K = \frac{LN - M^2}{EG - F^2}.$$

It is seen from this formula that the Gaussian curvature of a surface can be expressed in terms of the coefficients of the first and second fundamental forms. However, the Gaussian curvature can be expressed in terms of the coefficients and their derivatives of the first form alone (the Gauss theorem). Two surfaces between which it is possible to establish a homeomorphism preserving lengths of curves are called isometric surfaces.

The coefficients of the first and the second fundamental form are not independent. One relation between these coefficients is given by Gauss theorem. Two more relations have been discovered by K.M. Peterson and D. Codazzi. These three relations constitute a complete system of independent relations between the coefficients of the first and second fundamental forms of a surface. According to the Bonnet theorem, if the equations of Gauss, Peterson and Codazzi are satisfied for two differential fundamental forms the first one of which is positive definite, then there exists a surface, which is unique up to its location in space, with these forms as the first and second fundamental forms.

The branch of the theory of surfaces dealing with the properties on the surface which depend on a measurement of the lengths of curves on the surface alone is known as the interior geometry of surfaces. Since the lengths of curves are determined by the first fundamental form, the properties concerned are connected with that form only. In particular, the objects of the interior geometry of surfaces are the lengths of curves, the angles between curves, the areas of domains, and the Gaussian curvature. An important concept in the interior geometry of surfaces is the concept of a geodesic line. This is the name given to a curve which, over a sufficiently small part of the surface, is the shortest among all curves on the surface connecting the end points of this segment. The next important concept in the interior geometry of surfaces is that of the geodesic curvature of a curve. The Gauss–Bonnet theorem relates the integral of the Gaussian curvature of a surface over a domain, the integral of the geodesic curvature over its boundary and the Euler characteristic.

The interior geometry of surfaces may be constructed as the geometry of a two-dimensional metric manifold in which the distance between two points (u, v) and $(u+du, v+dv)$ which are infinitesimally close to each other is determined with the aid of a given differential form ds^2. If this approach to the interior geometry of surfaces is adopted, it can be naturally extended: the given manifold has arbitrary dimension n, while the metric is specified by a positive-definite differential quadratic form in n variables, $ds^2 = g_{\alpha\beta} du^\alpha du^\beta$. In a further generalization, the form ds^2 need not be positive definite. This leads to the theory of spaces of interest in the general theory of relativity, in particular, to Minkowski spaces. Finally, if the quadratic form ds^2 is replaced by a general positive homogeneous form of the first degree in du^α, one obtains a Finsler space. An even further-going generalization of the interior geometry of surfaces is the geometry of spaces with connections with respect to a given group, in particular the geometry of spaces with an affine connection, a projective connection and a conformal connection.

Euclidean Geometry

Euclidean geometry is the study of plane and solid on the basis of axioms and theorems employed by the Greek mathematician Euclid. In its rough outline, Euclidean geometry is the plane and solid geometry commonly taught in secondary schools. Indeed, until the second half of the 19th century, when non-Euclidean geometries attracted the attention of mathematicians, geometry meant Euclidean geometry. It is the most typical expression of general mathematical thinking. Rather than the memorization of simple algorithms to solve equations by rote, it demands true insight into the subject, clever ideas for applying theorems in special situations, an ability to generalize from known facts, and an insistence on the importance of proof. In Euclid's great work, the Elements, the only tools employed for geometrical constructions were the ruler and the compass—a restriction retained in elementary Euclidean geometry to this day.

In its rigorous deductive organization, the Elements remained the very model of scientific exposition until the end of the 19th century, when the German mathematician David Hilbert wrote his famous Foundations of Geometry. The modern version of Euclidean geometry is the theory of Euclidean (coordinate) spaces of multiple dimensions, where distance is measured by a suitable generalization of the Pythagorean Theorem.

Fundamentals

Euclid realized that a rigorous development of geometry must start with the foundations. Hence, he began the Elements with some undefined terms, such as "a point is that which has no part" and "a line is a length without breadth." Proceeding from these terms, he defined further ideas such as angles, circles, triangles, and various other polygons and figures. For example, an angle was defined as the inclination of two straight lines, and a circle was a plane figure consisting of all points that have a fixed distance (radius) from a given centre.

As a basis for further logical deductions, Euclid proposed five common notions, such as "things equal to the same thing are equal," and five unprovable but intuitive principles known variously as postulates or axioms. Stated in modern terms, the axioms are as follows:

- Given two points, there is a straight line that joins them.
- A straight line segment can be prolonged indefinitely.
- A circle can be constructed when a point for its centre and a distance for its radius are given.
- All right angles are equal.
- If a straight line falling on two straight lines makes the interior angles on the same side less than two right angles, the two straight lines, if produced indefinitely, will meet on that side on which the angles are less than the two right angles.

The fifth axiom became known as the "parallel postulate," since it provided a basis for the uniqueness of parallel lines. (It also attracted great interest because it seemed less intuitive or self-evident than the others. In the 19th century, Carl Friedrich Gauss, János Bolyai, and Nikolay Lobachevsky all began to experiment with this postulate, eventually arriving at new, non-Euclidean, and geometries). All five axioms provided the basis for numerous provable statements, or theorems, on which Euclid built his geometry.

Plane Geometry

Congruence of Triangles

Two triangles are said to be congruent if one can be exactly superimposed on the other by a rigid motion, and the congruence theorems specify the conditions under which this can occur. The first such theorem is the side-angle-side (SAS) theorem: If two sides and the included angle of one triangle are equal to two sides and the included angle of another triangle, the triangles are congruent. Following this, there are corresponding angle-side-angle (ASA) and side-side-side (SSS) theorems.

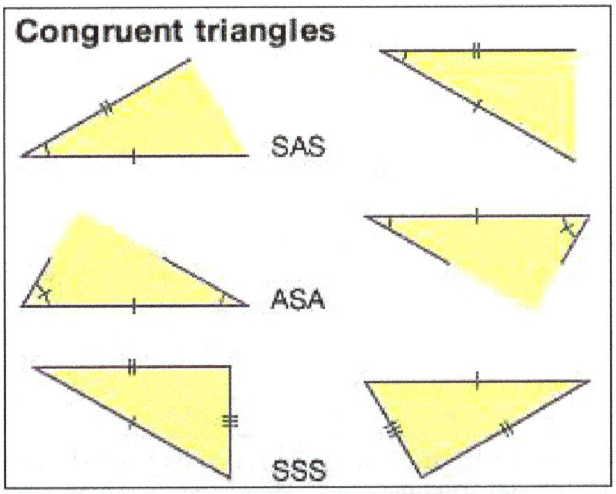

Congruent triangles: The figure above illustrates the three basic theorems that triangles are congruent (of equal shape and size) if: two sides and the included angle are equal (SAS); two angles and the included side are equal (ASA); or all three sides are equal (SSS).

The first very useful theorem derived from the axioms is the basic symmetry property of isosceles triangles—i.e., that two sides of a triangle are equal if and only if the angles opposite them are equal. Euclid's proof of this theorem was once called Pons Asinorum ("Bridge of Asses"), supposedly because mediocre students could not proceed across it to the farther reaches of geometry. The Bridge of Asses opens the way to various theorems on the congruence of triangles.

The parallel postulate is fundamental for the proof of the theorem that the sum of the angles of a triangle is always 180 degrees. A simple proof of this theorem was attributed to the Pythagoreans.

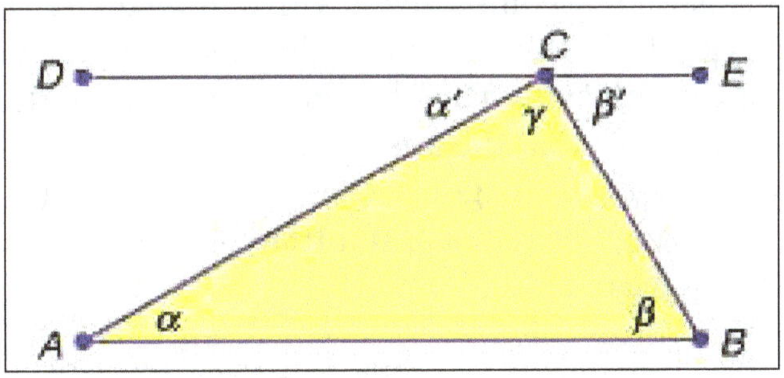

Proof that the sum of the angles in a triangle is 180 degrees: According to an ancient theorem, a transversal through two parallel lines (DE and AB in the figure) forms several equal angles, such as the alternating angles α/α' and β/β', labeled in the figure. By definition, the three angles α', γ, and β' on the line DE must sum to 180 degrees. Since $\alpha = \alpha'$ and $\beta = \beta'$, the sum of the angles in the triangle (α, β, and γ) is also 180 degrees.

Similarity of Triangles

Congruent figures have the same shape and size. Similar figures, on the other hand, have the same shape but may differ in size. Shape is intimately related to the notion of proportion, as ancient Egyptian artisans observed long ago. Segments of lengths *a*, b, c, and d are said to be proportional if *a*:b = c:d (read, *a* is to b as c is to d; in older notation *a*:b::c:d). The fundamental theorem of similarity states that a line segment splits two sides of a triangle into proportional segments if and only if the segment is parallel to the triangle's third side.

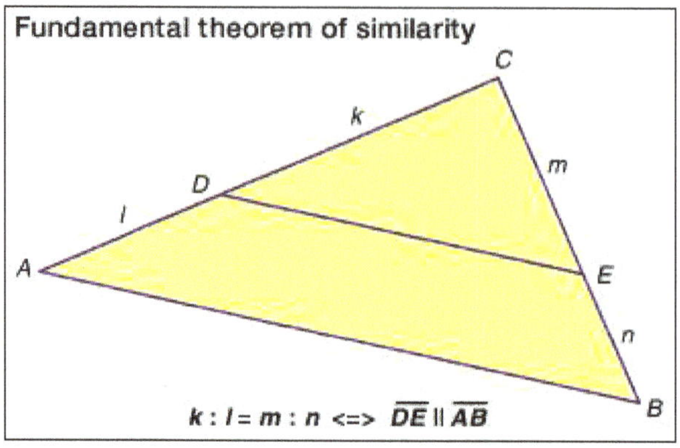

Fundamental theorem of similarity: The formula in the figure reads k is to l as m is to n if and only if line DE is parallel to line AB. This theorem then enables one to show that the small and large triangles are similar.

The similarity theorem may be reformulated as the AAA (angle-angle-angle) similarity theorem: two triangles have their corresponding angles equal if and only if their corresponding sides are proportional. Two similar triangles are related by a scaling (or similarity) factor s: if the first triangle has sides *a*, b, and c, then the second one will have sides sa, sb, and sc. In addition to the ubiquitous use of scaling factors on construction plans and geographic maps, similarity is fundamental to trigonometry.

Areas

Just as a segment can be measured by comparing it with a unit segment, the area of a polygon or other plane figure can be measured by comparing it with a unit square. The common formulas for calculating areas reduce this kind of measurement to the measurement of certain suitable lengths. The simplest case is a rectangle with sides *a* and b, which has area ab. By putting a triangle into an appropriate rectangle, one can show that the area of the triangle is half the product of the length of one of its bases and its corresponding height—bh/2. One can then compute the area of a general polygon by dissecting it into triangular regions. If a triangle has area A, a similar triangle with a scaling factor of s will have an area of s^2A.

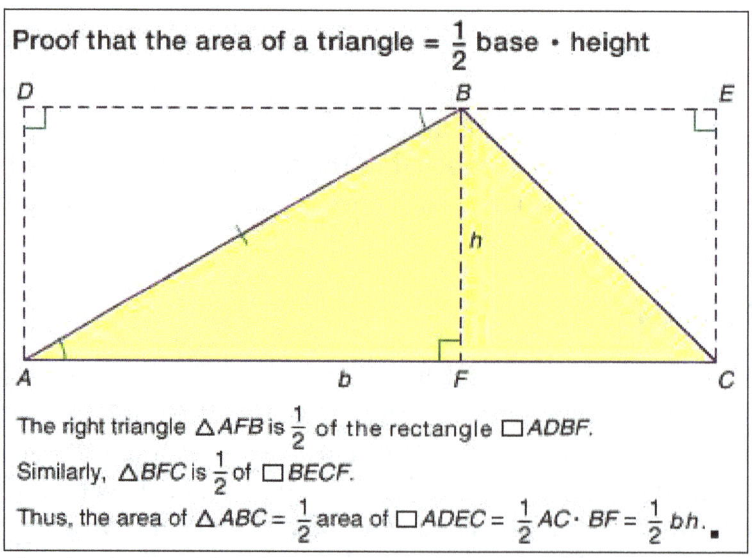

Area of a triangle: Proof that the area of a triangle is ½ the base times the height.

Pythagorean Theorem

For a triangle △ABC the Pythagorean Theorem has two parts: (1) if ∠ACB is a right angle, then $a^2 + b^2 = c^2$; (2) if $a^2 + b^2 = c^2$, then ∠ACB is a right angle. For an arbitrary triangle, the Pythagorean Theorem is generalized to the law of cosines: $a^2 + b^2 = c^2 - 2ab \cos(\angle ACB)$. When ∠ACB is 90 degrees, this reduces to the Pythagorean Theorem because $\cos(90°) = 0$.

Since Euclid, a host of professional and amateur have found more than 300 distinct proofs of the Pythagorean Theorem. Despite its antiquity, it remains one of the most important theorems in mathematics. It enables one to calculate distances or, more important, to define distances in situations far more general than elementary geometry. For example, it has been generalized to multidimensional vector spaces.

Circles

A chord AB is a segment in the interior of a circle connecting two points (A and B) on the circumference. When a chord passes through the circle's centre, it is a diameter, d. The circumference of a circle is given by πd, or $2\pi r$ where r is the radius of the circle; the area of a circle is πr^2. In each case, π is the same constant (3.14159...). The Greek mathematician Archimedes used the method of exhaustion to obtain upper and lower bounds for π by circumscribing and inscribing regular polygons about a circle.

A semicircle has its end points on a diameter of a circle. Thales (flourished 6th century BCE) is generally credited with having proved that any angle inscribed in a semicircle is a right angle; that is, for any point C on the semicircle with diameter AB, ∠ACB will always be 90°. Another important theorem states that for any chord AB in a circle, the angle subtended by any point on the same semiarc of the circle will be invariant. Slightly modified, this means that in a circle, equal chords determine equal angles, and vice versa.

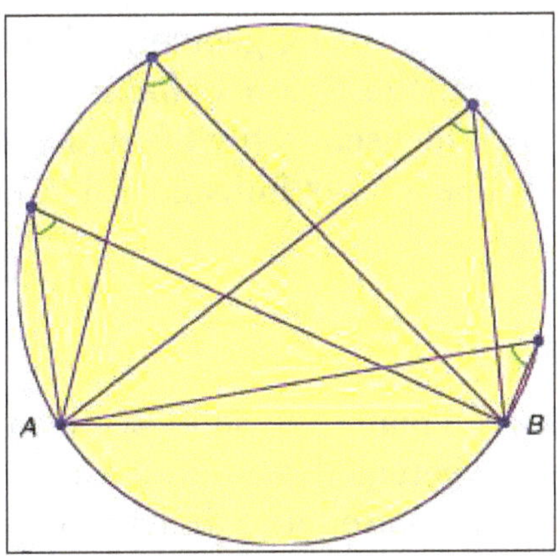

Chord in a circle: Thales of Miletus is generally credited with giving the first proof that for any chord AB in a circle, all of the angles subtended by points anywhere on the same semiarc of the circle will be equal.

Summarizing the above material, the five most important theorems of plane Euclidean geometry are: the sum of the angles in a triangle is 180 degrees, the Bridge of Asses, the fundamental theorem of similarity, the Pythagorean Theorem, and the invariance of angles subtended by a chord in a circle. Most of the more advanced theorems of plane Euclidean geometry are proved with the help of these theorems.

Regular Polygons

A polygon is called regular if it has equal sides and angles. Thus, a regular triangle is an equilateral triangle, and a regular quadrilateral is a square. A general problem since antiquity has been the problem of constructing a regular n-gon, for different n, with only ruler and compass. For example, Euclid constructed a regular pentagon by applying the five important theorems in an ingenious combination.

Techniques, such as bisecting the angles of known constructions, exist for constructing regular n-gons for many values, but none is known for the general case. In 1797, following centuries without any progress, Gauss surprised the mathematical community by discovering a construction for the 17-gon. More generally, Gauss was able to show that for a prime number p, the regular p-gon is constructible if and only if p is a "Fermat prime": $p=F(k) = 2^{2^k}+1$. Because it is not known in general which F(k) are prime, the construction problem for regular n-gons is still open.

Three other unsolved construction problems from antiquity were finally settled in the 19th century by applying tools not available to the Greeks. Comparatively simple algebraic methods showed that it is not possible to trisect an angle with ruler and compass or to construct a cube with a volume double that of a given cube. Showing that it is not possible to square a circle (i.e., to construct a square equal in area to a given circle by the same means), however, demanded deeper insights into the nature of the number π.

Conic Sections and Geometric Art

The most advanced part of plane Euclidean geometry is the theory of the conic sections (the ellipse, the parabola, and the hyperbola). Much as the Elements displaced all other introductions to geometry, the Conics of Apollonius of Perga, known by his contemporaries as "the Great Geometer," was for many centuries the definitive treatise on the subject.

Medieval Islamic artists explored ways of using geometric figures for decoration. For example, the decorations of the Alhambra of Granada, Spain, demonstrate an understanding of all 17 of the different "Wallpaper groups" that can be used to tile the plane. In the 20th century, internationally renowned artists such as Josef Albers, Max Bill, and Sol LeWitt were inspired by motifs from Euclidean geometry.

Solid Geometry

The most important difference between plane and solid Euclidean geometry is that human beings can look at the plane "from above," whereas three-dimensional space cannot be looked at "from outside." Consequently, intuitive insights are more difficult to obtain for solid geometry than for plane geometry.

Some concepts, such as proportions and angles, remain unchanged from plane to solid geometry. For other familiar concepts, there exist analogies—most noticeably, volume for area and three-dimensional shapes for two-dimensional shapes (sphere for circle, tetrahedron for triangle, box for rectangle). However, the theory of tetrahedra is not nearly as rich as it is for triangles. Active research in higher-dimensional Euclidean geometry includes convexity and sphere packings and their applications in cryptology and crystallography.

Volume

In plane geometry the area of any polygon can be calculated by dissecting it into triangles. A similar procedure is not possible for solids. In 1901 the German mathematician Max Dehn showed that there exist a cube and a tetrahedron of equal volume that cannot be dissected and rearranged into each other. This means that calculus must be used to calculate volumes for even many simple solids such as pyramids.

Regular Solids

Regular polyhedra are the solid analogies to regular polygons in the plane. Regular polygons are defined as having equal (congruent) sides and angles. In analogy, a solid is called regular if its faces are congruent regular polygons and its polyhedral angles (angles at which the faces meet) are congruent. This concept has been generalized to higher-dimensional (coordinate) Euclidean spaces.

Whereas in the plane there exist (in theory) infinitely many regular polygons, in three-dimensional space there exist exactly five regular polyhedra. These are known as the Platonic solids: the tetrahedron, or pyramid, with 4 triangular faces; the cube, with 6 square faces; the octahedron, with 8 equilateral triangular faces; the dodecahedron, with 12 pentagonal faces; and the icosahedron, with 20 equilateral triangular faces.

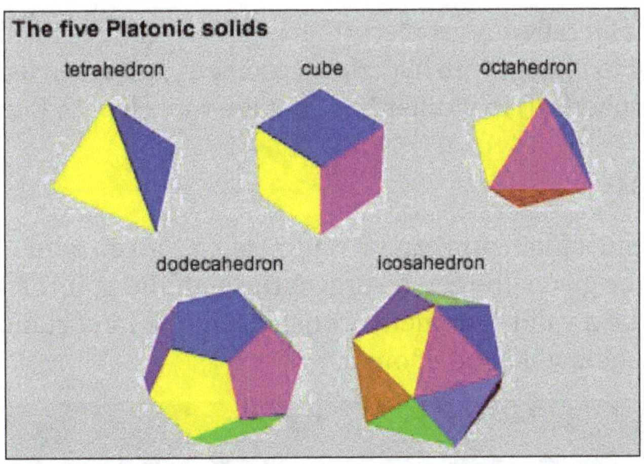

Platonic solids: These are the only geometric solids whose faces are composed of regular, identical polygons.

In four-dimensional space there exist exactly six regular polytopes, five of them generalizations from three-dimensional space. In any space of more than four dimensions, there exist exactly three regular polytopes—the generalizations of the tetrahedron, the cube, and the octahedron.

Calculating Areas and Volumes

The table presents mathematical formulas for calculating the areas of various plane figures and the volumes of various solid figures:

Mathematical formulas			
	Shape	Action	Formula
Circumference	Circle	Multiply diameter by π.	πd

Area	Circle	Multiply radius squared by π.	πr^2
	Rectangle	Multiply height by length	hl
	Sphere surface	Multiply radius squared by π by 4.	$4\pi r^2$
	Square	Length of one side squared.	S^2
	Trapezoid	Parallel side length A + parallel side length B multiplied by height and divided by 2.	$(A + B)h/2$
	Triangle	Multiply base by height and divide by 2.	$Hb/2$
Volume	Cone	Multiply base radius squared by π by height and divide by 3.	$br^2\pi h/3$
	Cube	Length of one edge cubed.	a^3
	Cylinder	Multiply base radius squared by π by height.	$br^2\pi h$
	Pyramid	Multiply base length by base width by height and divide by 3.	$lwh/3$
	Sphere	Multiply radius cubed by π by 4 and divide by 3.	$4\pi r^3/3$

Non-Euclidean Geometry

Non-Euclidean geometry is literally any geometry that is not the same as Euclidean geometry. Although the term is frequently used to refer only to hyperbolic geometry, common usage includes those few geometries (hyperbolic and spherical) that differ from but are very close to Euclidean geometry.

The Parallel Postulate

The fifth postulate in Euclid's Elements can be rephrased as, given a line and a point not on it, exactly one line parallel to the given line can be drawn through the point. The postulate is not true in 3D but in 2D it seems to be a valid statement. Considering the importance of postulates however, a seemingly valid statement is not good enough.

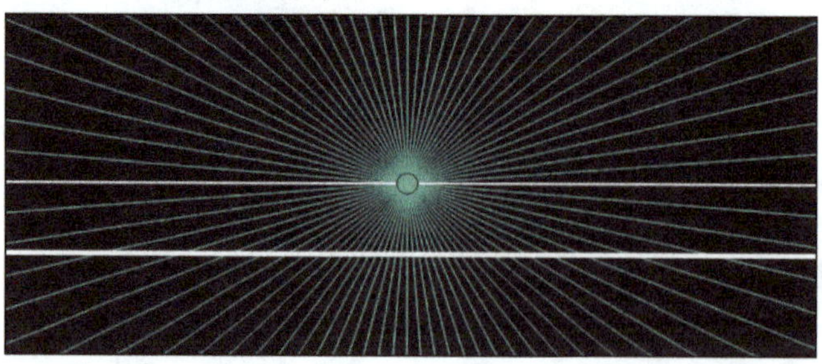

Exactly one line through the green point is parallel to the thick white line.

A postulate (or axiom) is a statement that acts as a starting point for a theory. Since a postulate is a starting point. As a statement that cannot be proven, a postulate should be self-evident. The Elements of Euclid is built upon five postulates. For over 2000 years the fifth postulate has been considered to be less intuitive than the other postulates, and not sufficiently self-evident. Many attempts have been made to prove the fifth postulate using the other four postulates. All these attempts have failed. In the 19th century it was shown that the fifth postulate is independent of the other postulates. It is possible to build a theory of geometry where the fifth postulate is not true. Such geometries are called non-Euclidean. Furthermore, it can be shown that non-Euclidean theories can be just as consistent as Euclidean theory. A theory is consistent if it does not contain any contradiction.

The converse statement to the parallel postulate is that there are either no lines or at least two lines "parallel to the given line through the point". It can be shown that if there are at least two lines, there are in fact infinitely many lines "parallel to."

If the parallel postulate is replaced by: Given a line and a point not on it, no lines parallel to the given line can be drawn through the point, you get an elliptic geometry. If the parallel postulate is replaced by: Given a line and a point not on it, infinitely many lines parallel to the given line can be drawn through the point, you get a hyperbolic geometry.

Distances, Angles and Lines

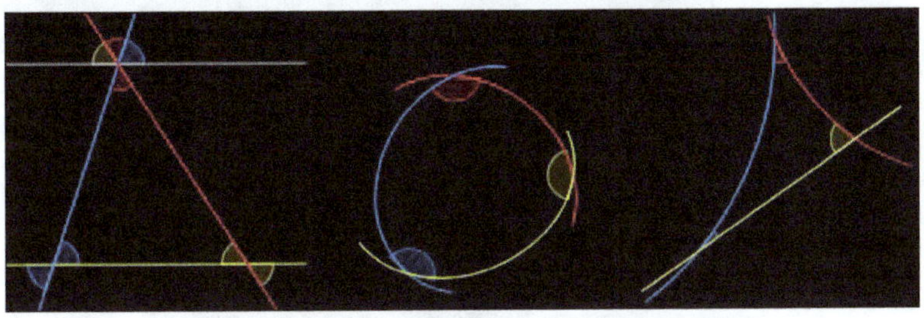

Euclidean, elliptical and hyperbolic triangle.

The theorems in the Elements of Euclid are built on a number of postulates and a number of definitions. Definition number 23 states that two lines are parallel if they never meet. There is nothing in the definition indicating that the distance between two parallel lines is the same everywhere. The parallel postulate is seemingly obvious only if you assume that parallel lines look like railroad tracks. If you redefine what you mean by a line, you may have that two parallel lines either converge towards each other or diverge from one another.

In non-Euclidean geometry, the concept corresponding to a line is a curve called a geodesic. In non-Euclidean geometry a shortest path between two points is along such a geodesic, or "non-Euclidean line". All theorems in Euclidean geometry that use the fifth postulate will be altered when you rephrase the parallel postulate. As an example; in Euclidean geometry the sum of the interior angles of a triangle is 180°, in non-Euclidean geometry this is not the case.

Model of Elliptic Geometry

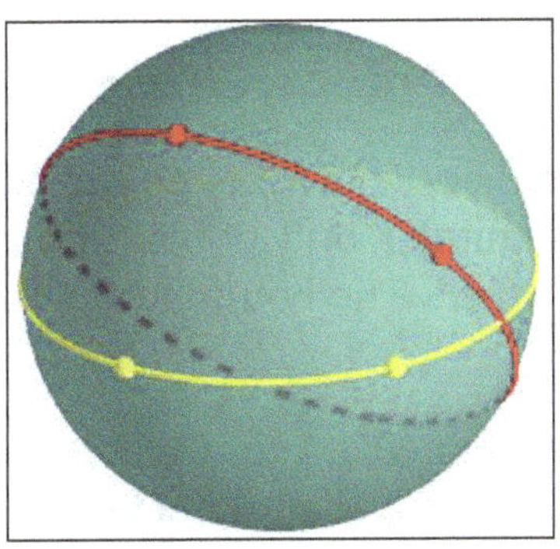

One easy way to model elliptical geometry is to consider the geometry on the surface of a sphere. A shortest path between two points on a sphere is along a so-called great circle. A great circle is a circle having the same radius, and the same centre, as the sphere. A great circle is a geodesic. If two points are directly opposite each other on the sphere (like the North Pole and South Pole) there are infinitely many shortest paths between the points. If the points are not antipodes, there is only one shortest path.

Two great circles that are not the same must intersect. Hence, there are no parallel lines on the surface of a sphere. A triangle is defined by three vertices and three arcs along great circles through each pair of vertices.

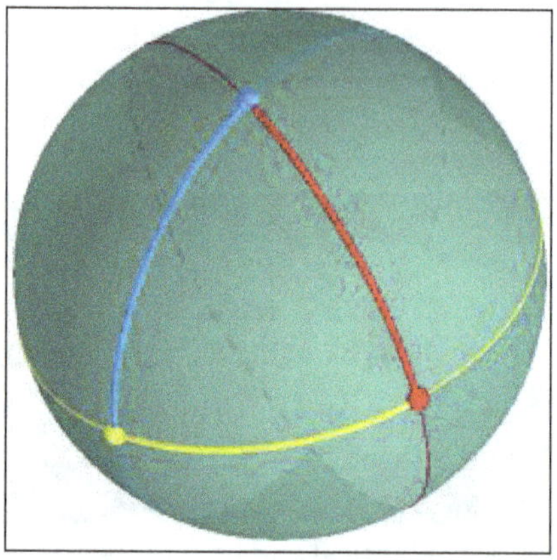

An interior angle at a vertex of a triangle can be measured on the tangent plane through that vertex. The sum of interior angles of an elliptical triangle is always > 180°.

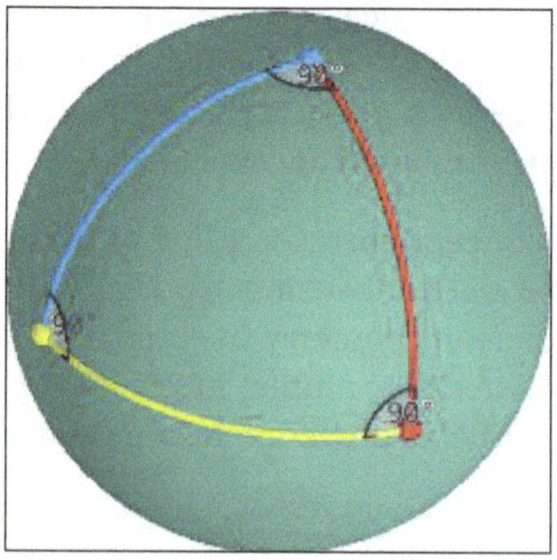

Hyperbolic Geometry using the Poincaré Disc Model

The Poincaré disc (in 2D) is an open disc, i.e. a set of points bounded by a circle not including the circle. When using the Poincaré disc model, only points in the Poincaré disc are considered. The Poincaré disc makes up the entire world. The boundary of the open disc is the circle at infinity, C_∞.

A geodesic through points A and B is defined as the circular arc through A and B that is perpendicular to C_∞. If A and B lie on a diameter of C_∞, that diameter is the geodesic through A and B.

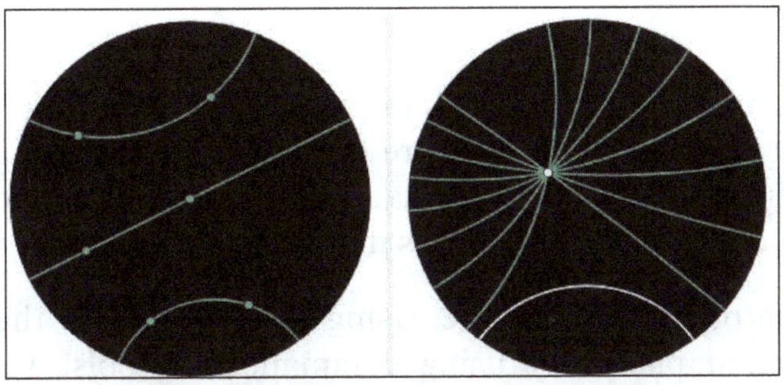

Geodesics defined by two points each: Given a white geodesic and a white point not on it, there are infinitely many geodesics through the white point not intersecting the white geodesic.

An interior angle of a triangle is measured between the corresponding tangent lines. The sum of interior angles is always < 180°. If the vertices of a triangle move towards circle C_∞, the angle sum tend to 0. Points that lie at C_∞ are called ideal points. The limiting triangle as the vertices tend to infinity (C_∞) is called an ideal triangle. An ideal triangle has angle sum 0. Since the vertices of an ideal triangle lie at infinity, the perimeter of an ideal triangle is infinite. Furthermore, it can be shown that the area of an ideal triangle is π. Hence, all the triangles in the topmost figure have the same hyperbolic area.

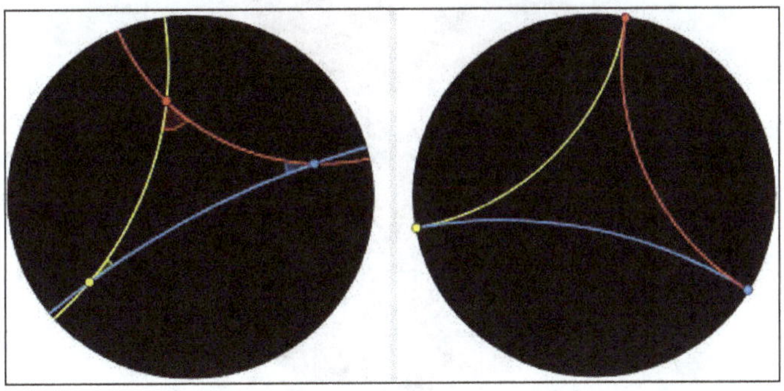

Hyperbolic triangles.

Hyperbolic Geometry using the Poincaré Halfplane Model

A line divides a plane into two halves. When using the Poincaré halfplane model, only points in half of a plane (not including the line) are considered. A geodesic through points A and B is defined as the circular arc through A and B that is perpendicular to the boundary line. If A and B lie on a line perpendicular to the boundary, a ray from the intersection point (not including the point) through A and B is the geodesic.

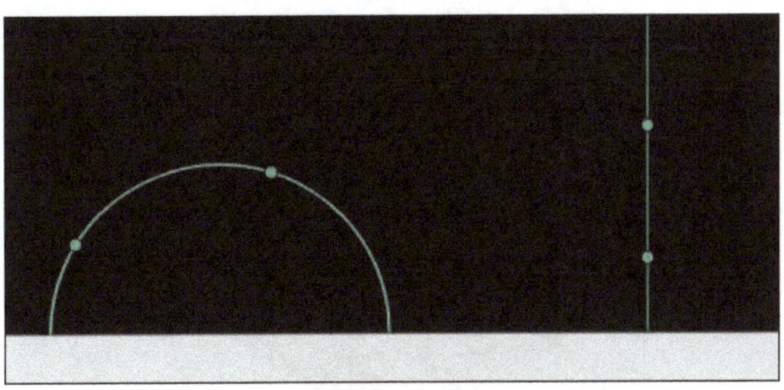

Geodesics using the half-plane model.

Exercise:

1. For non-Euclidean geometries all postulates in Euclid's Elements except the parallel postulate should be true. If you use geometry on a sphere to model elliptic geometry this is not the case (depending on how you interpret the first postulate). Use the Guide for Postulate 1 to explain why geometry on a sphere, as explained in the text, is not strictly non-Euclidean.

2. GeoGebra construction of elliptic geodesic. Using GeoGebra show the 3D Graphics win-dow. You can make spheres and planes by using commands or tools. To use a command start typing Sphere (or Plane) and GeoGebra will show you various alternatives for completing the code.

A circle on a sphere can be constructed by intersecting the sphere with a plane. You make an intersection by using the tool Intersect Two Surfaces. A great circle is constructed by choosing an appropriate plane to intersect the sphere with. Make a sphere and put two points A and B on the sphere. Construct the geodesic through A and B by using an appropriate plane. Hide all auxiliary objects.

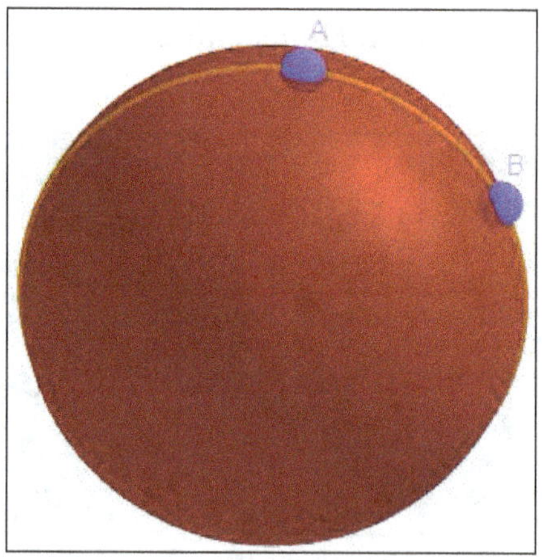

3. Measure an angle between two geodesics on a sphere using GeoGebra. Put three points A, B and C on a sphere and construct two geodesics that form the elliptic angle ∠ABC. The angle is measured by using points on the plane through B tangent to the sphere. Construct the tangent plane. Construct two points on the plane that can be used to measure the angle. Measure the angle and hide all auxiliary objects.

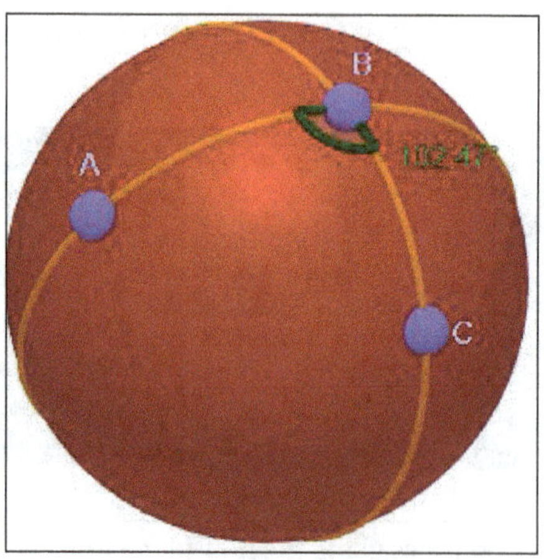

Algebraic Geometry

Algebraic geometry deals with curves or surfaces (or more abstract generalizations of these) which can be viewed both as geometric objects and as solutions of algebraic (specifically, polynomial) equations. A simple example is a circle of radius 1, which is the set of all points which are at a unit distance from its centre, but it is also the set of points (x, y) satisfying $x^2+y^2-1=0$. What geometric properties can be inferred from the equation? What about more complicated polynomial equations such as the semicubical parabola $y^2-x^3=0$, which has a cusp at its tip (such "singularities" are of great importance), or something less intelligible such as $x^2-x^2y^2+y^3+xy-1=0$? And, intriguingly, what if we change the minus sign in the equation for a circle to a plus, so that it reads $x^2+y^2+1=0$? Now x and y have to be complex, or there are no solutions at all. What does this imply?

Algebraic geometry sets out to answer these questions by applying the techniques of abstract algebra to the set of polynomials that define the curves (which are then called "algebraic varieties"). The mathematics involved is inevitably quite hard, although it is covered in degree-level courses. Other common questions in algebraic geometry concern points of special interest such as singularities, inflection points and points at infinity - we shall see these throughout the catalogue. More advanced questions in algebraic geometry concern relations between curves given by different equations and the topology of curves, and many other topics.

Algebraic geometry grew significantly in the 20th century, branching into topics such as computational algebraic geometry, Diophantine geometry, and analytic geometry. An area of particular significance here is singularity theory. We have many models which illustrate the classification of singularities on algebraic varieties.

Algebraic geometry is a very abstract subject, studied for beauty and interest alone. However, there are always interesting applications of pure mathematics, with algebraic geometry no. In algebraic statistics, techniques from algebraic geometry are used to advance research on topics such as the design of experiments and hypothesis testing. Another surprising application of algebraic geometry is to computational phylogenetic.

Singularities

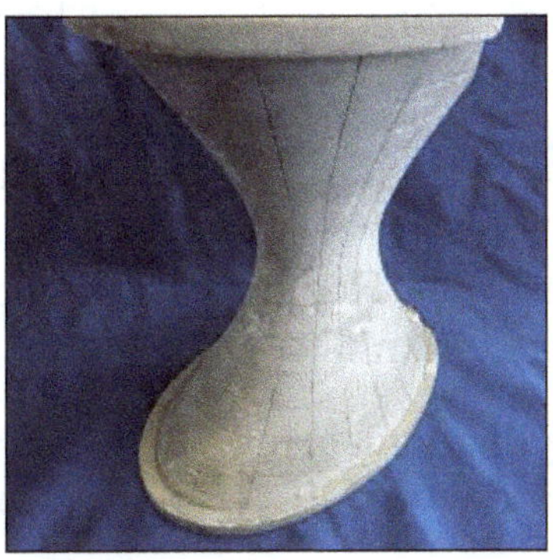

Consider a hyperboloid of one sheet, a cone, and a hyperboloid of two sheets. Imagine 'morphing'

between the one-sheet and two-sheet hyperboloids - the point at which we pass from one case to the other is the cone. On the cone, there is a singularity in the centre. Studying singularities is a crucial tool in geometry, as they occur at the 'boundary' of two different forms of behaviour. For example, the number of components of the surface changes here from one to two.

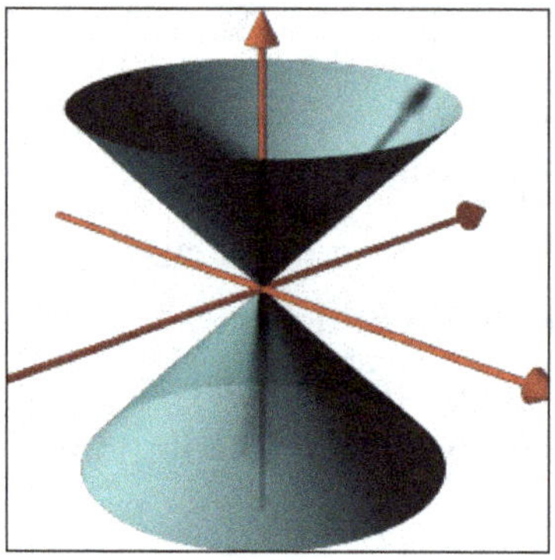

Surfaces defined by algebraic equations are "smooth" at almost all their points - by this we mean that, if you look closely enough, the surface looks like a plane ($\mathbb{R}^2$). A "singularity" (or "singular point") of an algebraic surface f(x, y, z) = 0 is a point at which it is not smooth, for example having a 'cusp' or a 'ridge'.

In a small neighbourhood of a singular point, the surface fails to resemble $\mathbb{R}^2$. This can be made more precise using topology: the surface does not locally resemble a manifold at singular points. It can also be helpful to look at just the real points on a surface, i.e. just those with coordinates in $\mathbb{R}$, satisfying the equation of the surface. We will speak of singularities as being "real" if they are singularities of the surface defined over the real and "complex" otherwise.

Rings, Ideals, and Polynomial Rings

Rings are a very powerful and useful object of study, especially within the topic of commutative algebra. A "ring" is a set together with two "operations", and can be thought of as a generalisation of the (positive and negative) integers with operations + and ×. Ring theory dates back to the 19th century, and has grown substantially with lots of mathematical topics, especially algebraic geometry, now relying on it.

An "ideal" is a special subset of a ring. An ideal is self-contained with respect to addition, and multiplying anything by a member of the ideal gives another member of the ideal. Intuitively, this generalises special subsets such as the even numbers within the integers. This term was coined by eminent German mathematician Richard Dedekind in the late 19th century, inspired by the notion of an ideal number.

A ring is "Noetherian" if all its ideals are finitely generated. This means that all members of an ideal can be found by "adding" and "multiplying", from some finite collection of starting elements. Working with such ideals is often desirable. Noetherian rings are named after Emmy Noether, who developed the foundation of modern commutative ring theory in 1921.

Given a ring R, its corresponding "polynomial ring" is the set of all 1 variable polynomials whose coefficients are members of R, and is denoted R[X]. For example, the integers $\mathbb{Z}$ have polynomial ring $\mathbb{Z}[X]$ containing elements such as x^2+3x-7 but not $x^3-\frac{1}{2}x+6$, which has a non-integer coefficient.

Algebraic Varieties and 2 Key Results

In algebraic geometry, the main objects of interest are "algebraic varieties", which are essentially geometric manifestations of solutions to polynomial equations. Some familiar examples include spheres, conic sections, and lines in $\mathbb{R}^2$ (two-dimensional Euclidean space). Algebraic surfaces can be thought of as particularly simple algebraic varieties, defined by only one polynomial equation. Our collection of mathematical models exhibits a range of interesting properties of algebraic surfaces.

Projective Space

A "field" is a special case of a ring, in which every non-zero element has a multiplicative inverse. This means that fields have a "division" operation analogous to what we have with real numbers. Indeed, the real numbers form a field. The integers do not, as $1/n$ is not an integer for $n \neq \pm 1$. (Note that when learning about new mathematical concepts, it is often very helpful to find such non-examples as well as examples of the concept itself).

A vector space (defined "over a field") is a set containing points, which we can add together and multiply by scalars (i.e. members of the field). This generalizes $\mathbb{R}^n$, familiar Euclidean space of dimension n, in which we can add coordinates together, and multiply (stretch) vectors by real numbers.

"Projective space", denoted P^n, is defined with respect to a field. It is perhaps best understood as a "quotient space", but can also be thought of as "the set of all lines through the origin" in a vector space (over a field). As a quotient space, for some $n + 1$-dimensional vector space V, projective space P^n is the set of all points in V equated up to scalar multiples, with the origin removed. This is the same as the set of lines through the origin. This gives rise to the notion of "homogeneous coordinates" (also called "projective coordinates"), which is the standard coordinate system for describing points in projective space. Many of our models represent surfaces in projective space, and are defined using homogeneous coordinates.

A relatively easy projective space to visualise is the projective plane P^2 which can be attained by taking all points on a sphere, and "gluing" antipodal points together. For a slightly more thorough introduction to projective space.

Dimension and Degree

An algebraic surface may also be thought of as an algebraic variety of dimension two. The notion of dimension of an algebraic variety generalizes the notions of dimension for the more familiar Euclidean space or a vector space. The dimension of a given variety can be defined as the greatest length d of a chain of distinct sub-varieties $X_0 \subset X_1 \subset \cdots \subset X_d$ all contained within each other. For example: a point, line, and plane can form a maximal chain with $l=2$, so a plane has dimension two, and is hence a (degenerate) algebraic surface. A complex cubic surface is an algebraic surface in P^3, over the field C of complex numbers. Recall the definition of the solutions to polynomial equations are considered over the complex numbers rather than the reals, and the polynomials involved can have complex coefficients).

An important property of an algebraic surface is its "degree" (also known as "order"), which may be thought of as the "total degree" of the polynomial defining it. Total degree refers to the sum of the exponents of variables. For example, the polynomial defined by $f(x,y,z)=x^3y-2y^2z+z^3$ has total degree four. For the more general case of an algebraic variety, the notion of degree is more complicated. It is defined using intersection theory. We have lots of models of algebraic surfaces of order 2, 3, and 4, respectively named quadric, cubic, and quartic surfaces. There are also a few surfaces of higher degree.

Complex Geometry

In plane geometry, complex numbers can be used to represent points, and thus other geometric objects as well such as lines, circles, and polygons. They are somewhat similar to Cartesian coordinates in the sense that they are used to algebraically prove geometric results, but they are especially useful in proving results involving circles and/or regular polygons (unlike Cartesian coordinates, which are useful for proving results involving lines).

Basics

A point in the plane can be represented by a complex number,

$$z=x+yi,$$

which corresponds to the Cartesian point (x, y). Therefore, the x-axis is renamed the real axis and the y-axis is renamed the imaginary axis, or imaginary line. This can also be converted into a polar coordinate (r, θ), which represents the complex number,

$$(r,\theta)=re^{i\theta},$$

which, intuitively speaking, means rotating the point (r,0) an angle of θ about the origin. By Euler's formula, this is equivalent to,

$$(r,\theta)=re^{i\theta}=r\cos\theta+ri\sin\theta,$$

which means that the polar coordinate (r, θ) corresponds to the Cartesian coordinate (r cos θ, r sin θ).

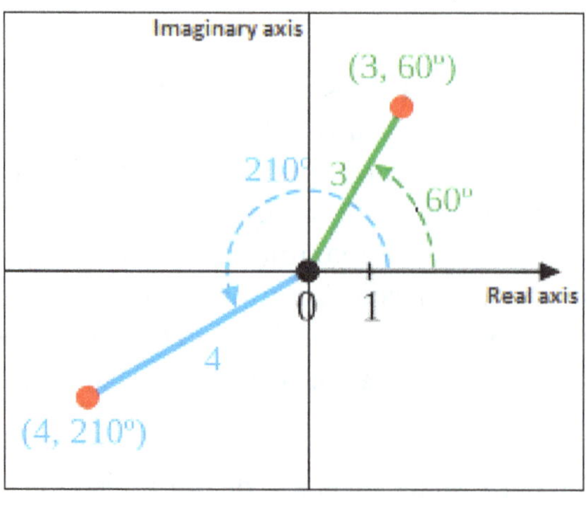

Incidentally, this immediately illustrates why complex numbers are so useful for circles and regular polygons: these involve heavy use of rotations, which are easily expressed using complex numbers.

In particular, a rotation of θ about the origin sends $z \to ze^{i\theta}$ for all θ.

In comparison, rotating Cartesian coordinates involves heavy calculation and (generally) an ugly result.

Additionally, each point $z = a + bi$ has an associated conjugate $\overline{z} = a - bi$. It satisfies the properties:

- $\overline{w+z} = \overline{w} + \overline{z}$
- $\overline{wz} = \overline{w} \cdot \overline{z}$
- $|z|^2 = z \cdot \overline{z}$

Geometrically, the conjugate can be thought of as the reflection over the real axis. This implies two useful facts: if z is real, $z = \overline{z}$, and if z is pure imaginary, $z = -\overline{z}$.

Lines in Complex Geometry

A capital letter denotes a point and the analogous lowercase letter denotes the complex number associated with it. Though lines are less nice in complex geometry than they are in coordinate geometry, they still have a nice characterization:

The points A, B, C are collinear if and only if $\frac{a-b}{b-c}$ is real, or equivalently, if and only if,

$$\frac{a-b}{\overline{a}-\overline{b}} = \frac{a-c}{\overline{a}-\overline{c}}.$$

There are two similar results involving lines. This is the one for parallel lines:

Lines AB and CD are parallel if and only if $\frac{a-b}{c-d}$ is real, or equivalently, if and only if,

$$\frac{a-b}{\overline{a}-\overline{b}} = \frac{c-d}{\overline{c}-\overline{d}}.$$

The following is the result for perpendicular lines:

Lines AB and CD are perpendicular if and only if $\frac{a-b}{c-d}$ is pure imaginary, or equivalently, if and only if,

$$\frac{a-b}{\overline{a}-\overline{b}} = -\frac{c-d}{\overline{c}-\overline{d}}.$$

Additionally, there is a nice expression of reflection and projection in complex numbers:

Let W be the reflection of Z over AB. Then,

$$w = \frac{(a-b)\overline{z} + \overline{a}b - a\overline{b}}{\overline{a}-\overline{b}}$$

and the projection of Z onto AB is $\frac{w+z}{2}$.

However, it is easy to express the intersection of two lines in Cartesian coordinates. In complex coordinates, this is not quite the case:

Lines AB and CD intersect at the point:

$$\frac{(\bar{a}b-a\bar{b})(c-d)-(a-b)(\bar{c}d-c\bar{d})}{(\bar{a}-\bar{b})(c-d)-(a-b)(\bar{c}-\bar{d})},$$

which is impractical to use in all but a few specific situations (e.g. when one of the points is at 0).

The Unit Circle

The unit circle is of special interest in the complex plane, as points z on the complex plane satisfy the key property that $z=\frac{1}{z}$, which is a consequence of the fact that $|z|=1$. This means that in general, complex geometry is most useful when there is a primary circle in the problem that can be set to the unit circle.

For instance, some of the formulas become significantly simpler. Reflection and projection, for instance, simplify nicely:

If A, B lie on the unit circle, the reflection of z across AB is $a+b-ab\bar{z}$. The projection of z onto AB is thus $\frac{1}{2}(z+a+b-ab\bar{z})$.

Also, the intersection formula becomes practical to use:

If A, B, C, D lie on the unit circle, lines AB and CD intersect at,

$$\frac{ab(c+d)-cd(a+b)}{ab-cd}.$$

Triangle Centers

Triangles in complex geometry are extremely nice when they can be placed on the unit circle; this is generally possible, by setting the triangle's circumcircle to the unit circle. This immediately implies the following obvious result:

Suppose A, B, C lie on the unit circle. Then the circumcenter of ABC is 0.

This is because the circumcenter of ABC coincides with the center of the unit circle.

The centroid is also easy to compute:

Suppose A, B, C lie on the unit circle. Then the centroid of A B C is $\frac{a+b+c}{3}$.

This also illustrates the similarities between complex numbers and vectors.

More interestingly, we have the following theorem:

Theorem: Suppose A, B, C lie on the unit circle. Then the orthocenter of ABC is a + b + c.

Proof, let h = a + b + c. AH is perpendicular to BC if and only if $\frac{h-a}{b-c}$ is pure imaginary. This is equal to $\frac{b+c}{b-c}$ since h = a + b + c. Then,

$$\overline{\left(\frac{b+c}{b-c}\right)}=\frac{\overline{b}+\overline{c}}{\overline{b}-\overline{c}}.$$

Since B, C are on the unit circle, $\overline{b}=\frac{1}{b}$ and $\overline{c}=\frac{1}{c}$. So,

$$\overline{\left(\frac{b+c}{b-c}\right)}=\frac{\overline{b}+\overline{c}}{\overline{b}-\overline{c}}=\frac{\frac{1}{b}+\frac{1}{c}}{\frac{1}{b}-\frac{1}{c}}=\frac{b+c}{c-b},$$

which implies $\overline{\left(\frac{b+c}{b-c}\right)}=-\left(\frac{b+c}{b-c}\right)$. This implies that $\left(\frac{b+c}{b-c}\right)$ is pure imaginary, so AH is perpendicular to BC.

By similar logic, BH is perpendicular to AC and CH to AB, so H is the orthocenter, as desired.

It is also possible to find the in center, though it is considerably more involved:

Suppose A, B, C lie on the unit circle, and let I be the in center. Let D, E, F be the feet of the angle bisectors from A, B, C, respectively. Then there exist complex numbers x, y, z such that $a=x^2, b=y^2, c=z^2, d=-yz, e=-xz, f=-xy$. Then,

$$I=-(xy+yz+zx).$$

Other Properties

There are two other properties worth noting before attempting some problems. The first is the tangent line through the unit circle:

Let W lie on the unit circle. Then Z lies on the tangent through W if and only if,

$$z+w^2\overline{z}=2w.$$

This is especially useful in the case of two tangents:

Theorem: Let X, Y be points on the unit circle. Their tangents meet at the point $\frac{2xy}{x+y}$, the harmonic mean of x and y.

Proof: Let Z be the intersection point. Then $z+x^2\overline{z}=2x$ and $z+y^2\overline{z}=2y$, so

$$\left(x^2-y^2\right)\overline{z}=2(x-y)\Rightarrow(x+y)\overline{z}=2\Rightarrow\overline{z}=\frac{2}{x+y}.$$

Thus, $z+\overline{\left(\frac{2}{x+y}\right)}=\frac{2}{\overline{x}+\overline{y}}$.

Since x, y lie on the unit circle, $\overline{x}=\frac{1}{x}$ and $\overline{y}=\frac{1}{y}$, so $z=\frac{2xy}{x+y}$, as desired.

The second result is a condition on cyclic quadrilaterals:

Points A, B, C, D lie on a circle if and only if $\dfrac{\frac{c-a}{c-b}}{\frac{d-a}{d-b}}$ is real.

Illustrative Problems

Example: A point A is taken inside a circle. For every chord of the circle passing through A, consider the intersection point of the two tangents at the endpoints of the chord. Find the locus of these intersection points.

WLOG assume that A is on the real axis. Let P, Q be the endpoints of a chord passing through A.

The tangents through p and q intersect at $z=\frac{2}{\overline{p}+\overline{q}}$. It is also true since P, A, Q are collinear, that,

$$\frac{p-a}{\overline{p}-\overline{a}}=\frac{a-q}{\overline{a}-\overline{q}}$$

$$\frac{p-a}{\frac{1}{p}-a}=\frac{a-q}{a-\frac{1}{q}}$$

$$pa-\frac{p}{q}+\frac{a}{q}=\frac{a}{p}-\frac{q}{p}+aq$$

$$p^2aq-p^2+ap=aq-q^2+apq^2$$

$$ap-aq+p^2aq-apq^2=p^2-q^2$$

$$a+apq=p+q$$

$$a=\frac{p+q}{pq+1}.$$

Now note that,

$$\text{Re } z=\frac{z+\overline{z}}{2}=\frac{1}{p+q}+\frac{1}{\overline{p}+\overline{q}}=\frac{pq+1}{p+q}=\frac{1}{a},$$

so z must lie on the vertical line through $\frac{1}{a}$.

For any point on this line, connecting the two tangents from the point to the unit circle at P and Q allows the above steps to be reversed, so every point on this line works; hence, the desired locus is this line. More formally, the locus is a line perpendicular to O A that is a $\frac{1}{OA}$ from O.

The following application of what we have learnt illustrates the fact that complex numbers are more than a tool to solve or "bash" geometry problems that have otherwise "beautiful" synthetic solutions, they often lead to the most beautiful and unexpected of solutions.

Theorem: Consider a polygonal line $P_0P_1...P_n$ such that $\angle P_0P_1P_2 = \angle P_1P_2P_3 = ... = \angle P_{n-2}P_{n-1}P_n$, all measured clockwise. If $P_0P_1 > P_1P_2 > ... > P_{n-1}P_n$, P_0 and P_n cannot coincide.

Proof: Let us consider complex coordinates with origin at P_0 and let the line P_0P_1 be the x-axis. Let α be the angle between any two consecutive segments and let $a_1 > a_2 > ... > a_n$ be the lengths of the segments. If we set $z = e^{i(\pi - \alpha)}$ then the coordinate of P_n We must prove that this number is not equal to zero.

Using the Abel Summation lemma, we obtain,

$$a_1 + a_2 z + ... + a_{n-1} z^n = (a_1 - a_2) + (a_2 - a_3)(1+z) + (a_3 - a_4)(1+z+z^2) + ... + a_n(1+z+...+z^{n-1}).$$

If α is zero, then this quantity is a strictly positive real number, and we are done.

If not, multiply by (1 - z) to get $(a_1 - a_2)(1-z)(a_2 - a_3)(1-z^2) + (a_3 - a_4)(1-z^3) + ... + a_n(1-z^n)$. This expression cannot be zero. Indeed, since $|z| = 1$, by the triangle inequality, we have,

$$\left|(a_1 - a_2)z + (a_2 - a_3)z^2 + (a_3 - a_4)z^3 + ... + a_n z^n\right| < (a_1 - a_2) + (a_2 - a_3) + (a_3 - a_4) + ... + a_n.$$

2

Fundamental Geometrical Concepts

Geometry has its own fundamental concepts which are crucial for gaining a better understanding of the subject. Axiom, point, line, line segments, plane, ray, angle, curve, manifold, length, area and volume are the fundamental geometrical concepts examined in detail in this chapter.

Axioms

Axioms, Conjectures and Theorems

Axioms or Postulate is defined as a statement that is accepted as true and correct, called as a theorem in mathematics. Axioms present itself as self-evident on which you can base any arguments or inference. These are universally accepted and general truth. 0 is a natural number, is an example of axiom.

The word 'Axiom' is derived from the Greek word 'Axioma' meaning 'true without needing a proof'. A mathematical statement which we assume to be true without a proof is called an axiom. Therefore, they are statements that are standalone and indisputable in their origins. In simpler words, these are truths that form the basis for all other derivations and have been derived from the basis of everyday experiences. In addition to this, there is no evidence opposing them.

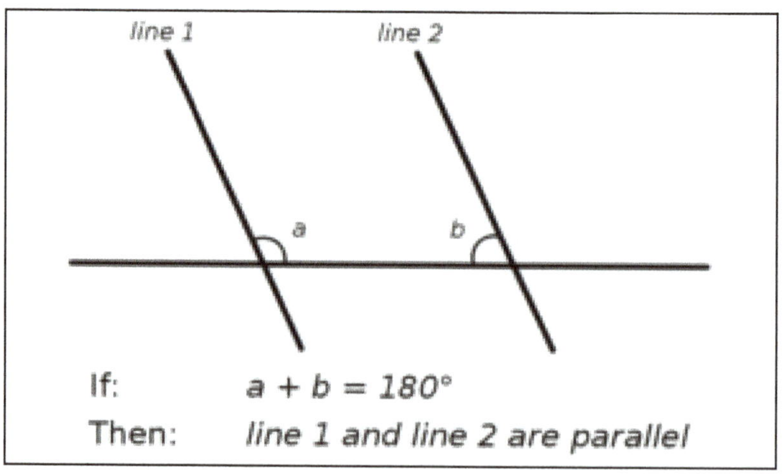

Examples of axioms can be 2 + 2 = 4, 3 x 3 = 4 etc. In geometry, we have a similar statement that a line can extend to infinity. This is an Axiom because you do not need a proof to state its truth as it is evident in itself.

Conjecture

A conjecture is such a mathematical statement whose truth or falsity we don't know yet. In other words, a statement that you believe to be true but have not proved to be true is called a conjecture. For example: What is the next number in the series 3 6 9 12? The answer is '15'. On careful observation, we see that each succeeding number is greater than the previous one by a difference of '3'. So even though we don't see the next number we can correctly guess it by observing the pattern generated.

So, one of the conjectures is that "the next number is 15". Another conjecture could be "the next number is (15 × 1) + 0". None of the two has a proof but both follow from simple mathematical rules or axioms. Conjectures play a very important role in problem-solving in Mathematics and Geometry, where the solution is not always apparent and we generate the solution by following a series of steps. Generally, each of these steps is a 'Conjecture' over the previous step.

Theorem

A mathematical statement that we know is true and which has a proof is a theorem. We can further explain it as a series of Conjectures (proof) that combine together to give a true result. So if a statement is always true and doesn't need proof, it is an axiom. If it needs a proof, it is a conjecture. A statement that has been proven by logical arguments based on axioms is a theorem. We generate a theorem by the way of analysis and proof. Consider them your weapons, Superheroes as you start on this journey through the mazes of mathematics and arrive at solutions and save the day.

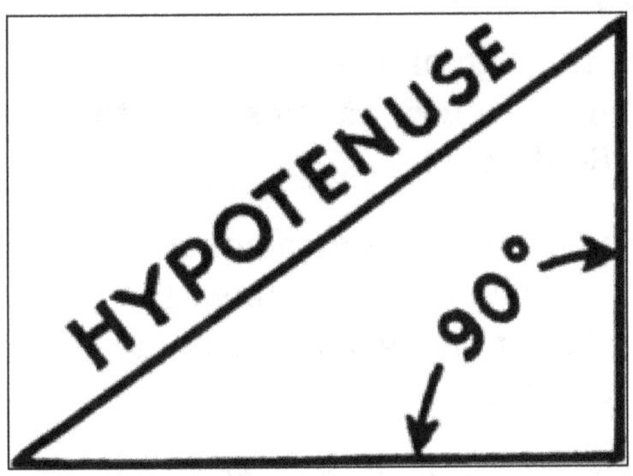

The Axioms of Euclidean Plane Geometry

For well over two thousand years, people had believed that only one geometry was possible, and they had accepted the idea that this geometry described reality. One of the greatest Greek achievements was setting up rules for plane geometry. This system consisted of a collection of undefined terms like point and line, and five axioms from which all other properties could be deduced by a formal process of logic. Four of the axioms were so self-evident that it would be unthinkable to call any system geometry unless it satisfied them:

- A straight line may be drawn between any two points.
- Any terminated straight line may be extended indefinitely.
- A circle may be drawn with any given point as center and any given radius.
- All right angles are equal.
- But the fifth axiom was a different sort of statement: If two straight lines in a plane are met

by another line, and if the sum of the internal angles on one side is less than two right angles, then the straight lines will meet if extended sufficiently on the side on which the sum of the angles is less than two right angles.

Because this axiom was much more complicated than the previous axioms, it seemed more like a theorem than a self-evident proposition. Since all attempts to deduce it from the first four axioms had failed, Euclid simply included it as an axiom because he knew he needed it. For example, some axiom like this one was necessary for proving one of Euclid's most famous theorems, that the sum of the angles of a triangle is 180 degrees. Mathematicians found alternate forms of the axiom that were easier to state, for example: For any given point not on a given line, there is exactly one line through the point that does not meet the given line.

This form of the fifth axiom became known as the parallel postulate. Although it was simpler to understand than Euclid's original formulation, it was no easier to deduce from the earlier axioms. The attempt to deduce the fifth axiom remained a great challenge right up to the nineteenth century, when it was proved that the fifth axiom did not follow from the first four.

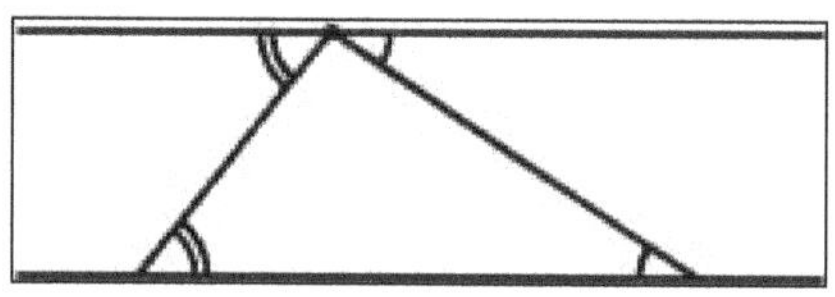

Once we can draw a unique line through one vertex of a triangle not meeting the line containing the opposite side, we can use alternate interior angles to see that the sum of the angles of a triangle is the same as a straight angle, 180 degrees.

The great advantage of expressing geometry as an axiomatic system was that it no longer was necessary to memorize long lists of independent facts about the nature of the universe--one only had to know a small set of axioms, and by applying to them the rules of inference, one could reconstruct the entire collection of geometric truths.

There was little doubt that the Greeks were attempting to describe a real world when they formulated their geometry, even though it might have been an ideal sort of world, realized only abstractly "in the mind of God." Many mathematicians, now as well as in the distant past, believe that the complete structure of mathematics is something that exists in itself and that it is only gradually discovered by human beings laboring to uncover its mysteries. Even though the framers of the early axiom systems would refer to point and line as undefined terms, they fairly clearly thought of them as real objects, and they thought that the system they were developing was a progressively more and more elaborate and accurate description of the real world. The progress of algebra, on the other hand, was not quite so settled, and people accepted changes in viewpoint there more readily than in the very traditional field of geometry.

Point

A point in geometry is a location. It has no size i.e. no width, no length and no depth. A point is shown by a dot.

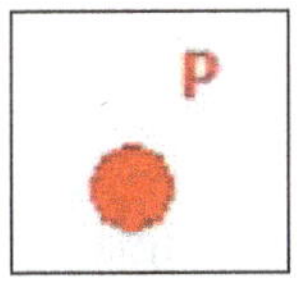

A line is defined as a line of points that extends infinitely in two directions. It has one dimension, length. Points that are on the same line are called collinear points.

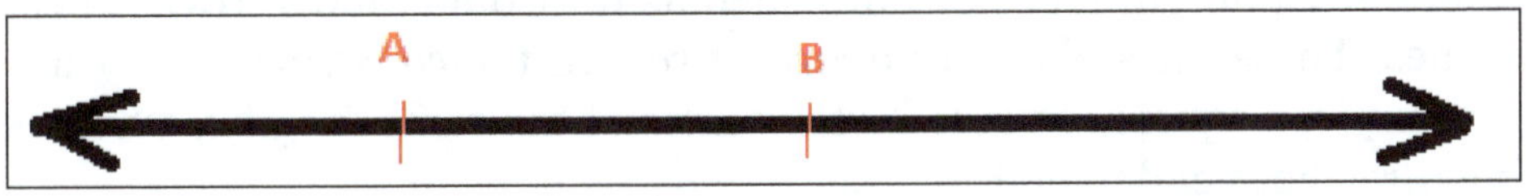

A line is defined by two points and is written as shown below with an arrowhead.

$\overleftrightarrow{AB}$

Two lines that meet in a point are called intersecting lines.

A part of a line that has defined endpoints is called a line segment. A line segment as the segment between A and B above is written as:

$\overline{AB}$

A plane extends infinitely in two dimensions. It has no thickness. An example of a plane is a co-ordinate plane. A plane is named by three points in the plane that are not on the same line. Here below we see the plane ABC.

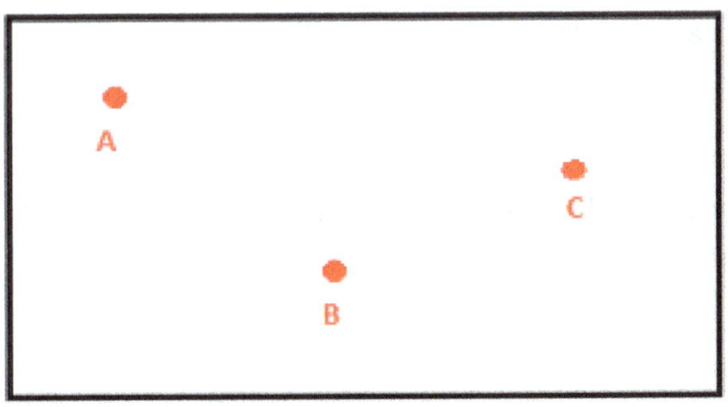

A space extends infinitely in all directions and is a set of all points in three dimensions. You can think of a space as the inside of a box.

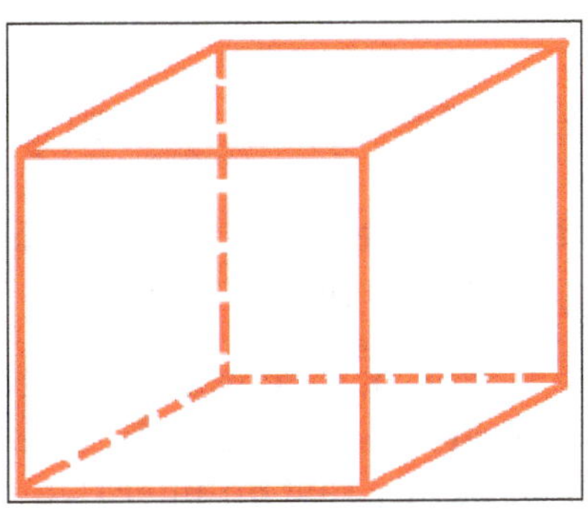

Line

A line is a one-dimensional figure, which has length but no width. A line is made of a set of points which is extended in opposite directions infinitely. It is determined by two points in a two-dimensional plane. The two points which lie on the same line are said to be collinear points.

In geometry, there are different types of lines such as horizontal and vertical lines, parallel and perpendicular lines. These lines play an important role in the construction of different types of polygons. For example, a square is made by four lines of the same lengths, whereas a triangle is made by joining three lines end to end.

A line can be defined as a straight set of points that extend in opposite directions:

- It has no ends in both directions (infinite).
- It has no thickness.
- It is one-dimensional.

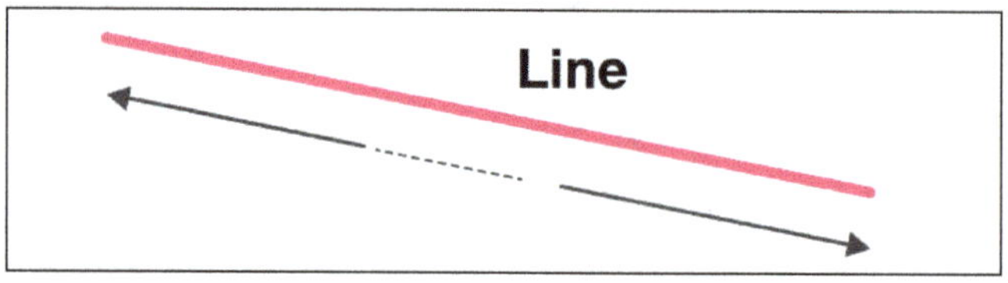

Types of Line

In Geometry, there are basically four types of lines. They are:

Horizontal Line

A horizontal line is a line that goes left-to-right or right-to-left. We often think of a horizontal line as being the opposite of a vertical line.

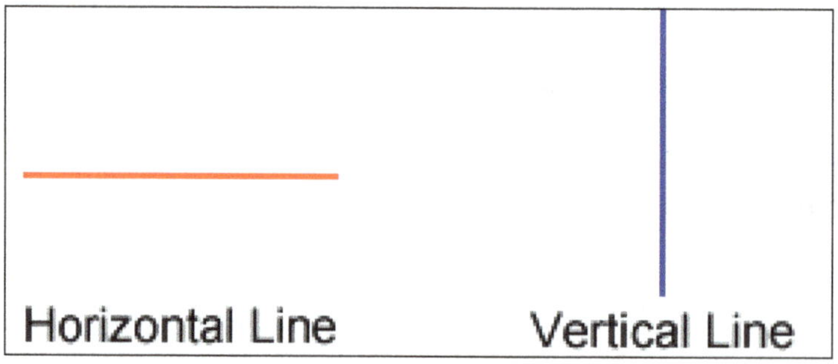

In coordinate geometry, a horizontal line is parallel to the x-axis on the coordinate plane. Below is the graph of a vertical line that has a y-intercept of (0, k).

The horizontal line is parallel to the x-axis so, it does not intersect it. Any horizontal line does not intersect the x-axis unless the horizontal line is the x-axis. The y-coordinate for any point on the vertical line is k. The x-coordinate can be any real number (we say x is arbitrary).

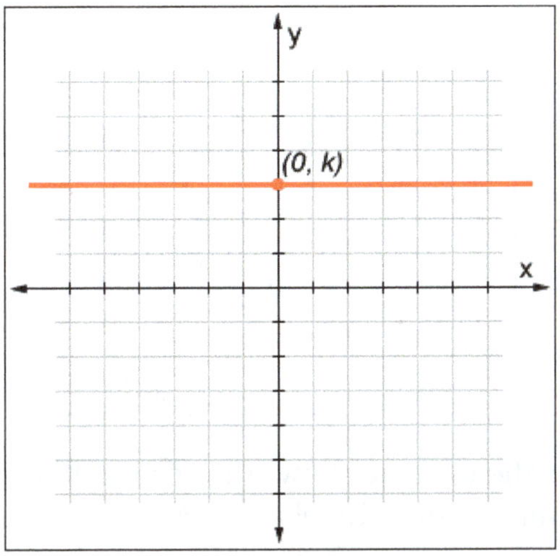

The slope for a horizontal line is zero. Using the points (0, -4) and (3, -4) for the horizontal line below, the slope is:

$$m = \frac{-4-(-4)}{0-3} = \frac{0}{-3} = 0$$

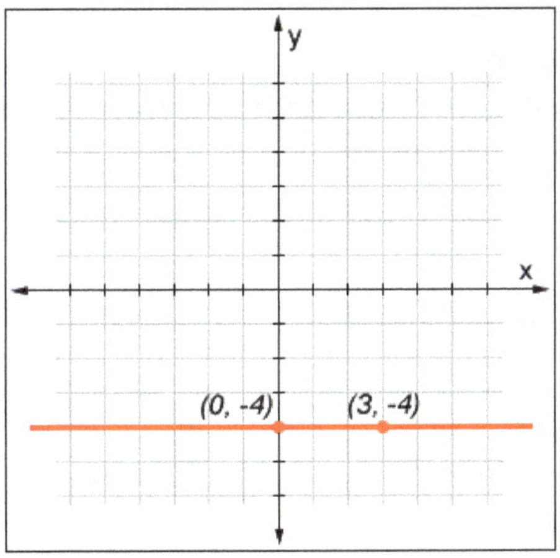

Since the slope is zero for a horizontal line, we can use the slope-intercept form, y = mx + b to write an equation for it. Since m = 0, it becomes y = b.

Vertical Line

In a coordinate plane, a line parallel to the Y-axis is called Vertical Line. It is a straight line which goes from top to bottom and bottom to top. Any point in this line will have the same value for the x-coordinate. For example, (2,0), (3,0) (-4,0), etc. are the points of vertical lines. Similarly, the line which goes from left to right and is parallel to the x-axis is called a horizontal line.

The vertical lines have no slope. It goes parallels along with y-axis and hence the slope is not defined. Therefore, the equation of the vertical line which crosses the x-axis at any point say 'a', is given by: x = a, where x is the coordinate of a point on the line and 'a' is the point where the lines cross the x-intercept.

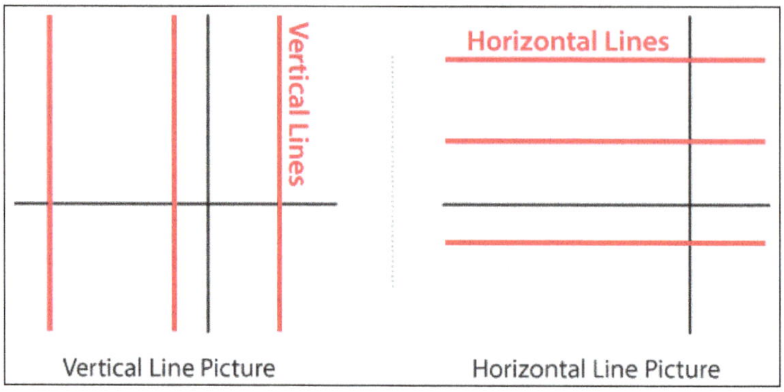

In the above figure, we can see, the left side shows the vertical lines in a coordinate plane whereas the right side shows the horizontal lines parallel to the x-axis.

Properties of Vertical Lines

- The equation of a vertical line in the graph, which is parallel to y-axis is x = a.
- The slope of a vertical line is infinity or undefined as it has no y-intercept and the denominator in the slope formula is zero.
- To check whether the relation is a function in math's we use a vertical line. It would be a function if all vertical lines intersect it minimum once. This is also called a vertical line test. A graph will be considered as a function if it has only one output y for each input x. Therefore, a vertical line cannot be a function.

Solved Examples

Question: Plot the graph for the following:

1) X = 8

x = 8 is a vertical line, that passes through x-axis, at a distance of 8 units on the right of the origin and is parallel to y-axis.

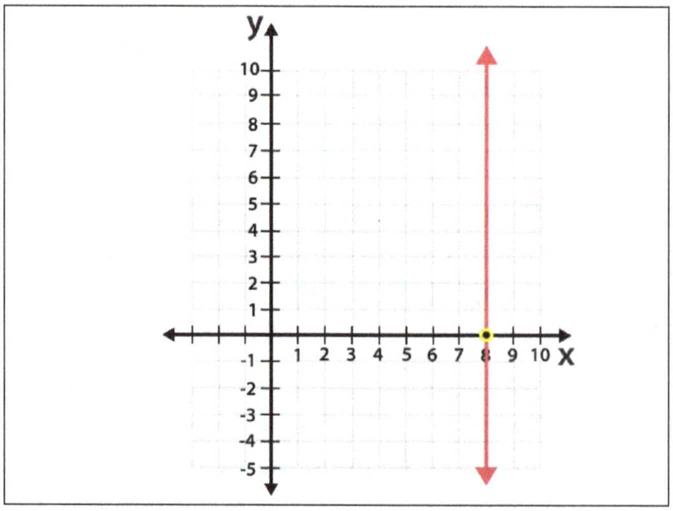

2) X = – 4

x = -4 is a vertical line, that passes through x-axis, at a distance of 4 units on the left side of the origin and is parallel to y-axis.

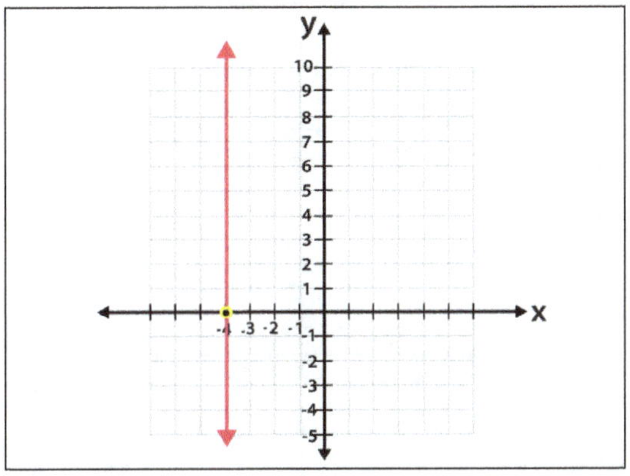

3) X = 3

x = 3 is a vertical line, that passes through x-axis, at a distance of 3 units on the right side of the origin and is parallel to y-axis.

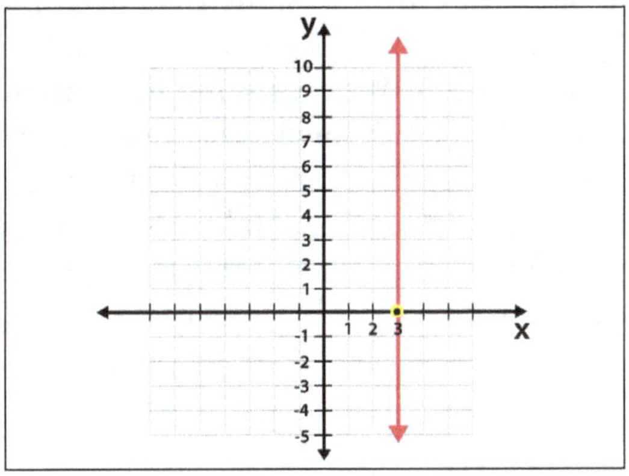

4) X = – 5

x = -5 is a vertical line, that passes through x-axis, at a distance of 5 units on the left of the origin and is parallel to y-axis.

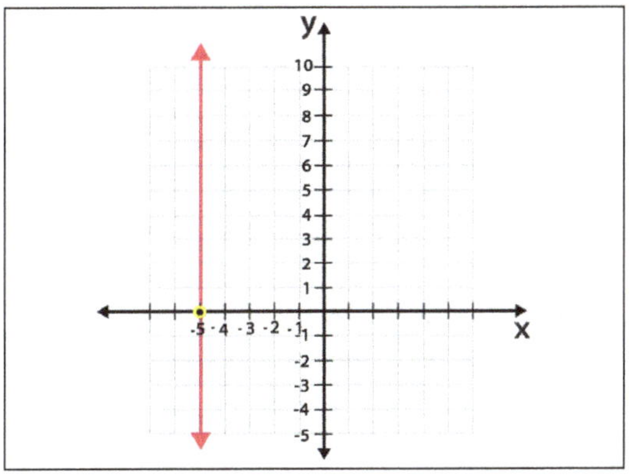

5) X = 6

x = 6 is a vertical line, that passes through x-axis, at a distance of 6 units on the right side of the origin and is parallel to y-axis.

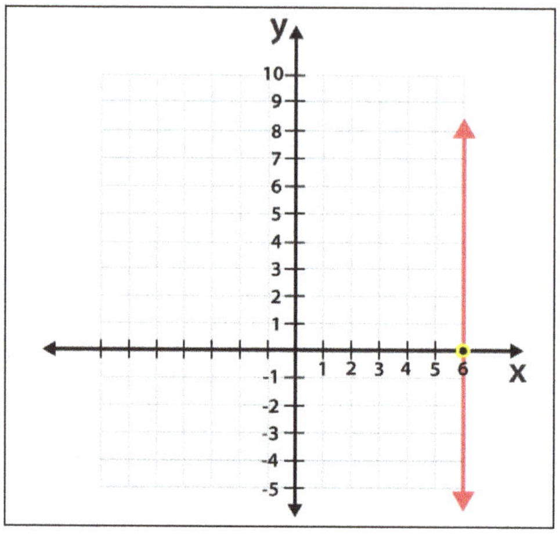

Parallel Line

Parallel lines are two or more lines that are the same distance apart, never merging and never diverging. The English word "parallel" is a gift to geometricians, because it has two parallel lines in it, in the form of the two side-by-sides is. We recall that lines are never-ending, so parallel lines continue forever in two directions. They don't have to be straight lines, as long as they are always the same distance apart. A single existing line can have an infinite number of parallel lines, too, simply by picking a point, not on any existing line, and constructing a line parallel to all the existing ones. To be parallel lines, both lines must exist in the same plane; they must be coplanar. This differentiates them from skew lines, which exist in three dimensions (two separate planes).

Parallel Shapes

Parallel lines are by themselves very interesting, but when you take segments of them, you can construct many polygons. Think of squares, parallelograms, rectangles, and trapezoids; they all have parallel sides made from parallel line segments.

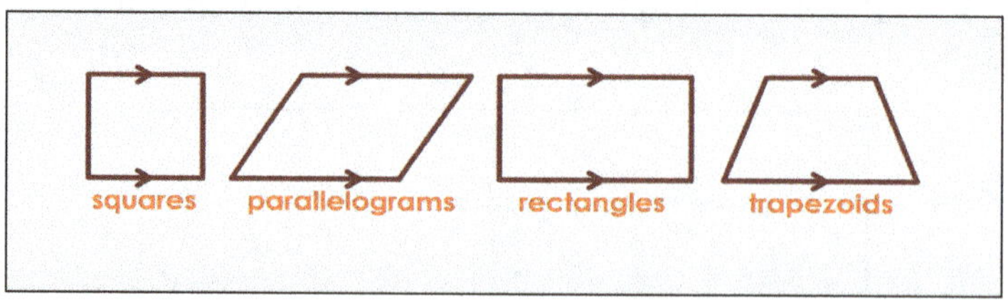

Parallel Symbol

To symbolize parallel lines in geometry, we use two vertical lines (or slightly slanted lines), like this:

AT || UP

AT / / UP

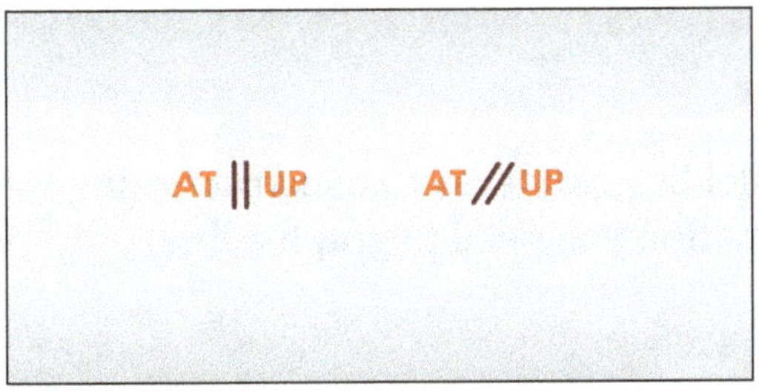

Both of those statements tell us that line AT is parallel to line UP.

How to Construct Parallel Lines

Construct parallel lines with a straightedge, a pencil, and plain paper. If using a ruler, this is a one-step process. Lay the ruler on the paper, holding it carefully, so it does not slip. Trace each edge of the ruler.

When you move the ruler, you have parallel lines. You can label them, identifying each line with two named points and placing arrowheads at the end of the two lines. You can also construct a line parallel to an existing line passing through a single point, not on the line. You need a drawing compass plus the plain paper, pencil, and straightedge.

- For the given line, draw a transversal crossing the existing line and passing through the point not on the line; we'll call that Point A. Now, you have two intersecting lines. Identify the point where your new transversal intersects the existing line; call it Point N.
- Set your drawing compass to scribe an arc shorter than the distance from the existing line to Point A.
- Scribe two arcs with the compass needle set on Point N and again on Point A. The first arc, from Point N, should cross the existing line and the transversal; the second arc, from Point A, will cross only the transversal somewhere above Point A.
- Where your first arc crosses the transversal below Point A, place your compass needle on the transversal and adjust the compass to reach the point on the existing line where your first arc crossed it.
- Lift the compass, and without adjusting it, place the needle on that point above Point A where the second arc intersected the transversal.
- Scribe another arc down to cross that second arc.

- Where the two arcs cross near Point A, connect that point to Point A to construct a line parallel to your existing line.

Graphing Parallel Lines

In coordinate graphing, parallel lines are easy to construct using the grid system. A simple equation can provide all the information you need to graph a line:

$$3x - y = -4$$

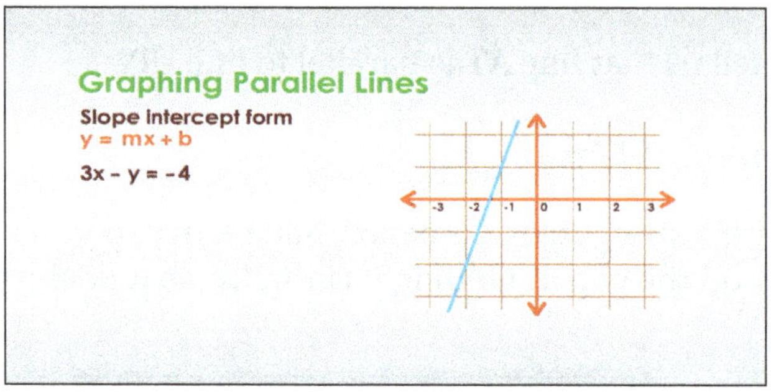

Our line is established with the slope-intercept form, $y = mx + b$. So we solve the first equation, so it is changed to the slope-intercept form:

$$3x - y = -4$$

Subtract 3x from both sides:

$$3x - y - 3x = -4 - 3x$$

Simplify:

$$-y = -4 - 3x$$

Divide both sides by -1:

$$\frac{-y}{-1} = \frac{-4}{-1} - \frac{-3x}{-1}$$

$$y = 3x + 4$$

Now add another line that keeps the x-value the same. You only need to retain 3x, so you could write:

$$y = 3x - 5$$

This new line is far to the right of the first line but is exactly parallel because the x-value determined its slope. Same x-value; same slope; both lines parallel.

Real-Life Examples of Parallel Lines

Parallel lines rule the construction industry. Buildings are erected with walls parallel to each other, ceilings are parallel to floors, and one building is usually put up parallel to other buildings on the same block. Notebook paper is a thick collection of parallel lines. When your notebook is closed, the lines are not only parallel on each page; they are parallel from page to page, too.

Parallel lines as a purely geometric concept must be completely flat and exist on a single plane. In real life, while railroad tracks, the edges of sidewalks, and the markings on streets are all parallel, the tracks, sidewalks, and streets go up and down hills and around curves. Those three real-life examples are good, but not perfect, models of parallel lines.

Consider railroad tracks: The two lines never meet, but they move up and down and side to side through three-dimensional space. Railroad tracks, unless they are on a completely flat plain like a desert, are more similar to skew lines than parallel lines.

Perpendicular Line

Perpendicular lines in math are two lines where the angle between them is 90°. The symbol used in between the two lines to say that they are perpendicular is ⊥. If two lines $\overline{AB}$ and $\overline{CD}$ are perpendicular to each other, we represent it using the notation $\overline{AB} \perp \overline{CD}$.

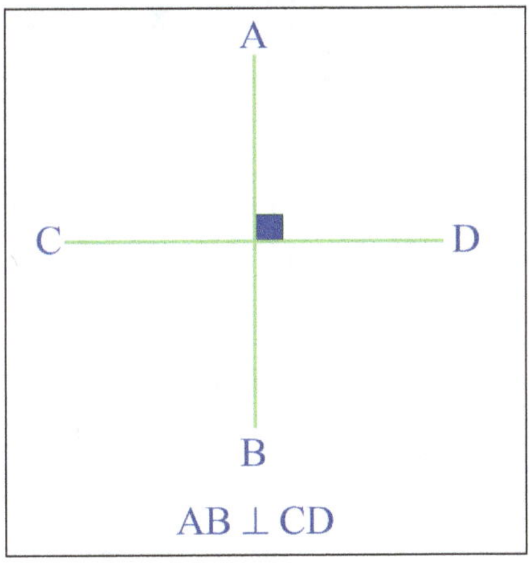

Perpendicular shapes in geometry refer to shapes that include perpendicular lines.

Some perpendicular shapes in geometry are:

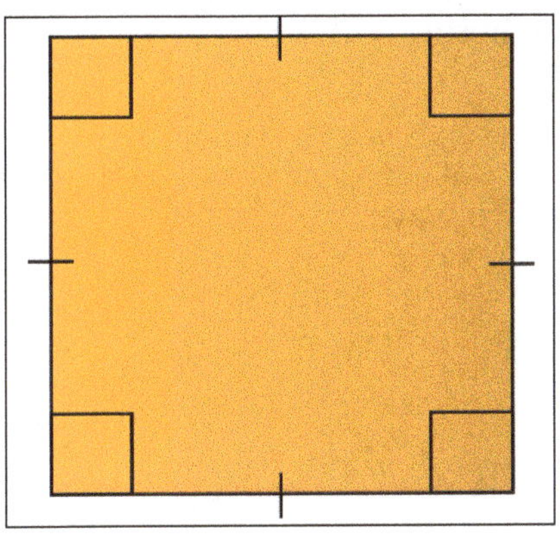

Square

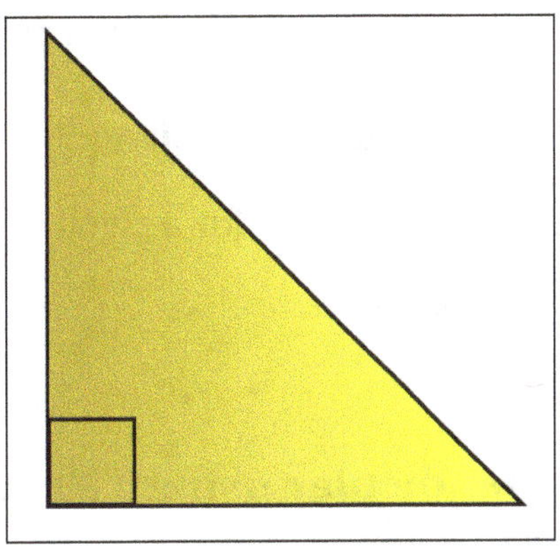

Right-angled triangle

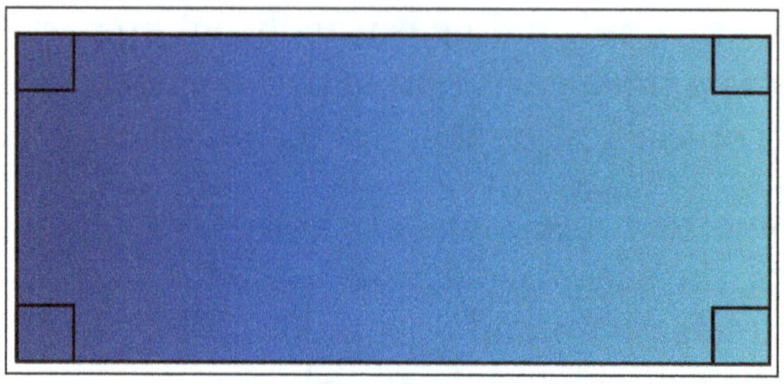

Rectangle

Here are some real-life examples of perpendicular lines:

Whiteboard

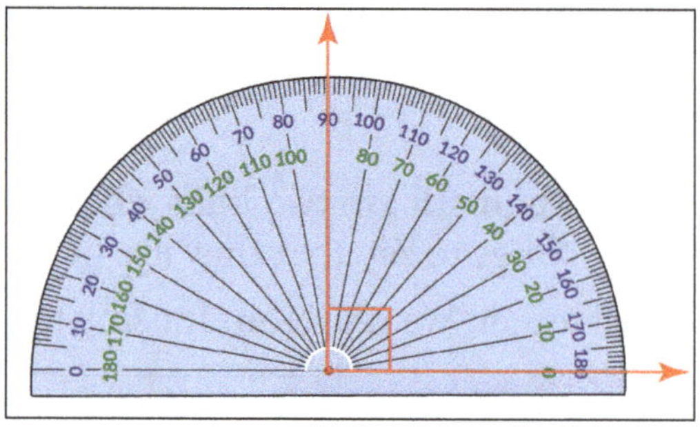

Protractor

Clock

You can understand more about perpendicular lines using the following simulation. Here, you can drag the line O A by holding it at A and you can see whether $OA \perp OB$.

How to Draw Perpendicular Lines?

We can draw a perpendicular line for a given line in two ways:

- Using protractor,
- Using compass.

Drawing a Perpendicular using a Protractor

To draw a perpendicular line at a point P on a line.

Step 1: Place the baseline of the protractor along the line such that its centre is at P.

Step 2: Mark a point or a dot B at 90° of the protractor.

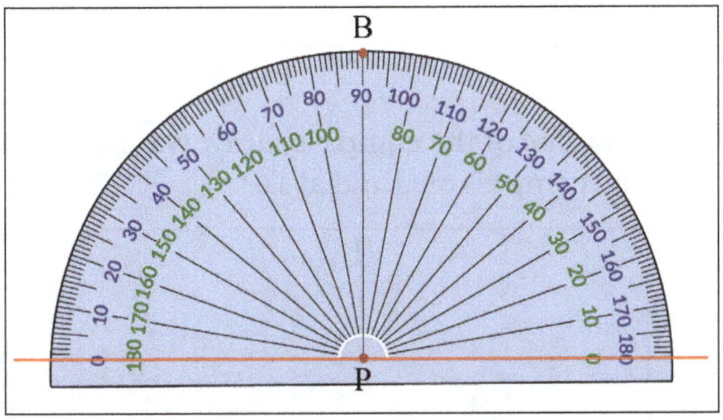

Step 3: Remove the protractor and join P and B. This forms a perpendicular line for the given line.

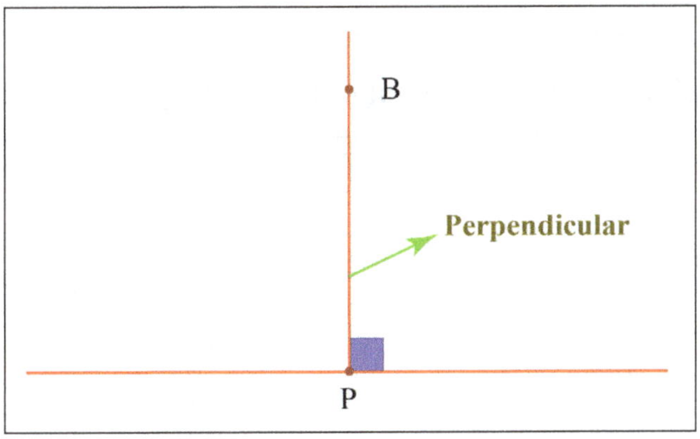

Drawing a Perpendicular using a Compass

We can draw a perpendicular line without using a protractor but by using a compass instead. Let's see how. To draw a perpendicular line P at a point:

Step 1: Open the compass to a desired radius.

Step 2: Place the pointer of the compass at P and draw a semi-circle which cuts the line at A and B.

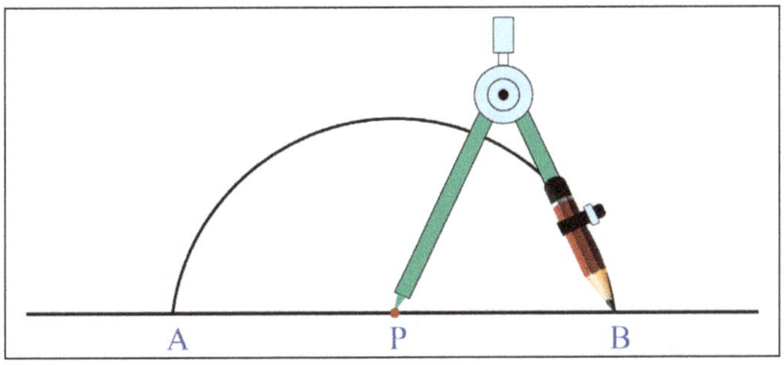

Step 3: Without disturbing the radius of the compass, draw two arcs which cut the semi-circle at C and D by placing the pointer of the compass at A and at B.

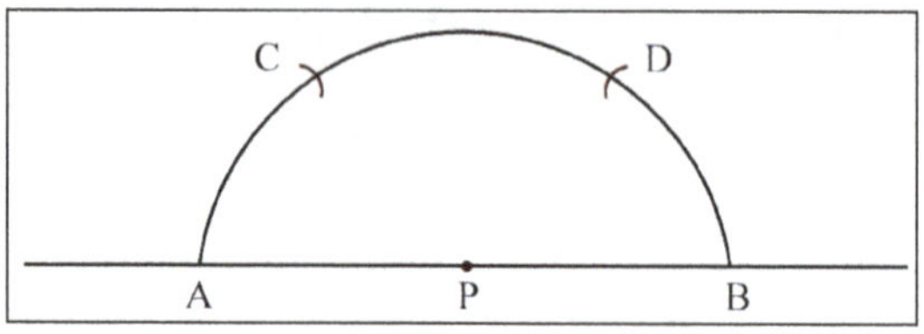

Step 4: Without disturbing the radius of the compass, draw two intersecting arcs which intersect at Q by placing the pointer of the compass at C and at D.

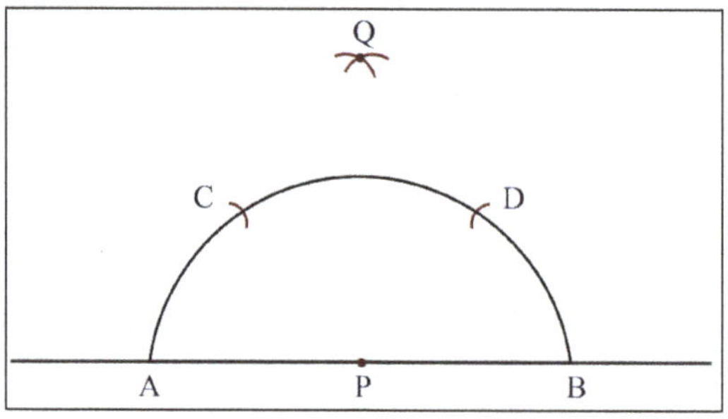

Step 5: Join P and Q which forms a perpendicular line of the given line.

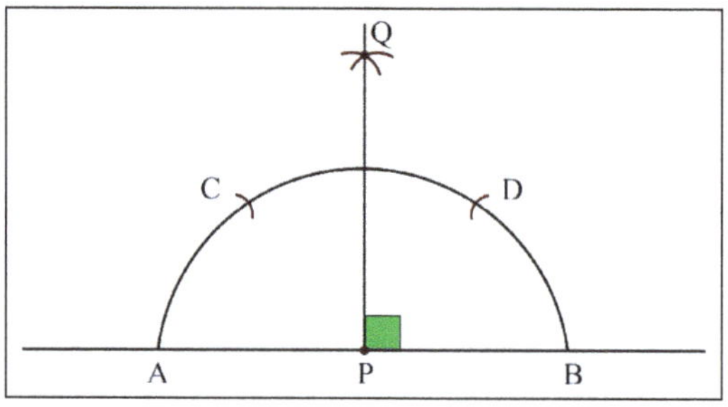

Properties of Perpendicular Lines

There are two important properties of perpendicular lines in math:

- The angle between any two perpendicular lines is always equal to 90°.
- If two lines are perpendicular to the same line, the two lines are parallel.

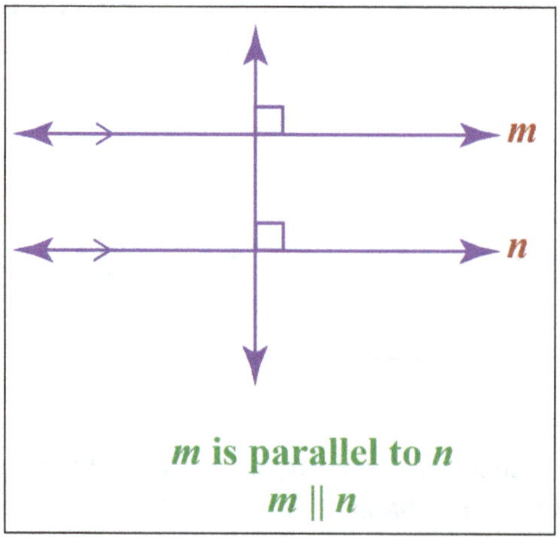

Some other Types of Lines in Maths

Tangent Lines

The Tangent is a straight line which just touches the curve at a given point. The normal is a straight line which is perpendicular to the tangent. To calculate the equations of these lines, we shall make use of the fact that the equation of a straight line passing through the point with coordinates (x_1, y_1) and having gradient m is given by,

$$\frac{y-y_1}{x-x_1}=1$$

We also make use of the fact that if two lines with gradients m_1 and m_2 respectively are perpendicular, then $m_1m_2 = -1$.

Example: Suppose we wish to find points on the curve y(x) given by x^3-6x^2+x+3 where the tangents are parallel to the line y = x + 5.

Solution: If the tangents have to be parallel to the line then they must have the same gradient. The standard equation for a straight line is y = mx + c, where m is the gradient. So what we gain from looking at this standard equation and comparing it with the straight line y = x + 5 is that the gradient, m, is equal to 1. Thus, the gradients of the tangents we are trying to find must also have gradient 1.

We know that if we differentiate y(x) we will obtain an expression for the gradients of the tangents to y(x) and we can set this equal to 1. Differentiating and making it equal to 1 we find:

$$\frac{dy}{dx}=3x^2-12x+1=3$$

from which,

$$3x^2-12x=0$$

This is a quadratic equation which we can solve by factorization.

$$3x^2-12x=0$$
$$3x(x-4)=0$$
$$\Rightarrow 3x=0 \text{ or } (x-4)=0$$
$$\Rightarrow x=0 \text{ or } x=4$$

Now having found these two values of x we can calculate the corresponding y coordinates. We do this from the equation of the curve: $y=x^3-6x^2+x+3$.

$$\text{when } x=0:\ y=(0)^3-6(0)^2+0+3=3$$

$$\text{when } x=4:\ y=(4)^3-6(4)^2+4+3=-25$$

So the two points are (0, 3) and (4, −25).

These are the two points where the gradients of the tangent are equal to 1, and so where the tangents are parallel to the line that we started out with, i.e. $y = x + 5$.

Secant Lines

A line in the plane is a secant line to a circle if it meets the circle in exactly two points. It is also equivalent to the average rate of change, or simply the slope between two points. The average rate of change of a function between two points and the slope between two points are the same thing.

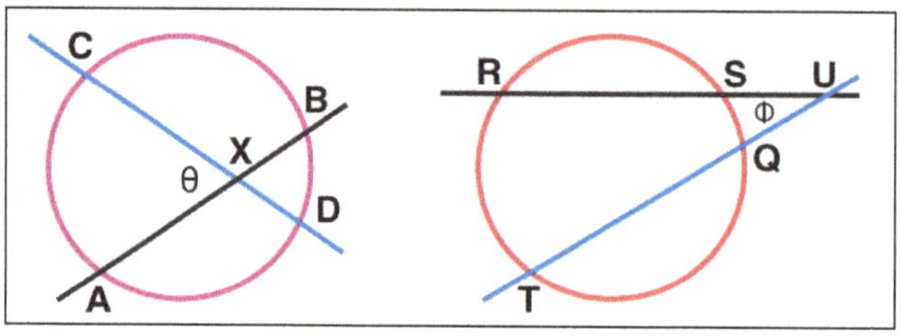

In the left figure above,

$$\theta = \frac{1}{2}(\text{arc } A\,C + \text{arc } B\,D),$$

while in the right figure,

$$\phi = \frac{1}{2}(\text{arc } R\,T - \text{arc } S\,Q),$$

Where arc AB denotes the angular measure of the arc AB.

Ray

A ray can be thought of as being a snippet or segment of a line. In plane geometry, a ray is easily constructed with two points. One will be an endpoint, the start of the ray. The other point is merely a signpost, a way to give the ray a name. The line that connects the two points extends in only one direction infinitely:

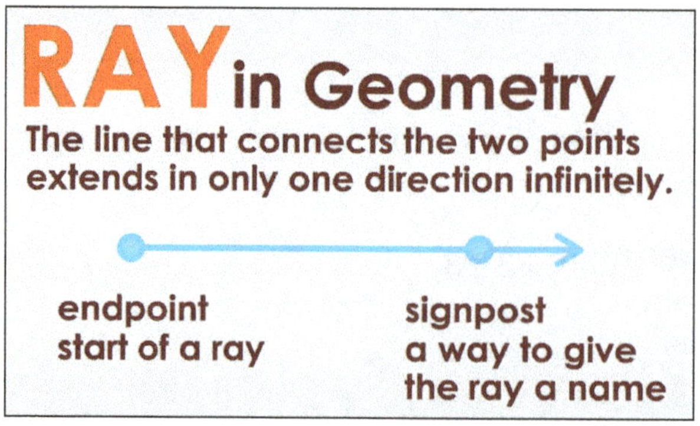

Instead of allowing both ends of the line to go on forever, we snip one side at a given point. Now we have a ray.

How to Draw a Ray in Math

To draw a ray, place two points on a piece of paper. Label both points with capital letters. Choose one point to be the endpoint. Use a straightedge to draw a line starting at your endpoint and continuing through your second point. Draw one arrowhead on the open end of your line (the one opposite the endpoint).

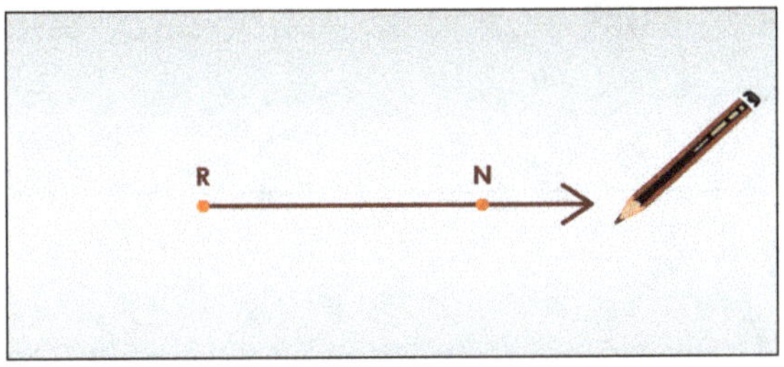

Ray Symbol and Label

To symbolize and label a ray, we need that endpoint identified. We also need some other point along the one-way line. Then, we write the endpoint and other point together as capital letters, capped by a tiny, one-way arrow (pointing to the right):

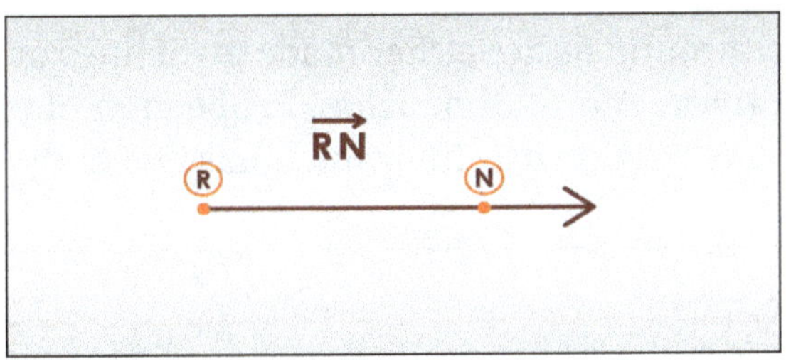

This is the symbol for Ray $\overrightarrow{RN}$ named after an NFL quarterback, who can throw a football that very nearly moves like a ray. Gravity tugs the football down, but the quarterbacks' arm speed and strength can make short passes look like straight-line rays. He is the endpoint; the traveling football is the one-way line.

Ray in Geometry Examples

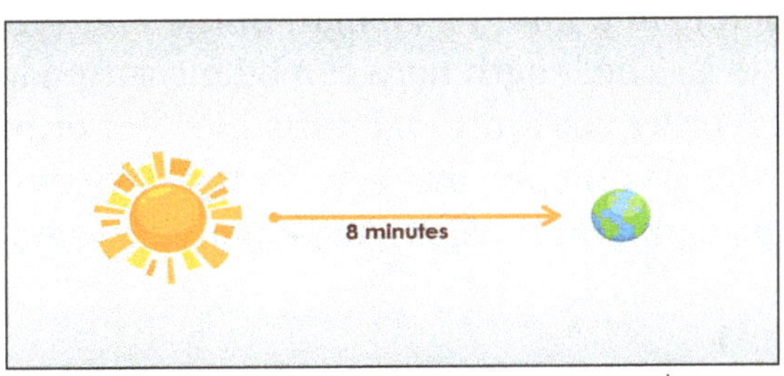

A ray of sunshine is a ray. It originates at our star, the Sun, and travels one way, striking earth some eight minutes after it left its "endpoint," the Sun. Tennis pro, Rafael Nadal, famously serves tennis balls at some 217 kph (135 mph), which defies gravity's tug so well it seems to travel in a straight line, just like a ray. The light beam from a classroom LCD projector is a ray; so is light from a movie projector at your local cinema. The path an arrow travels from a bow is a ray and has the added benefit of being, well, arrow-shaped.

Lasers are excellent examples of rays because unlike sports balls, they are not much affected by earth's gravity, so they shine in steady, straight one-way lines from their source.

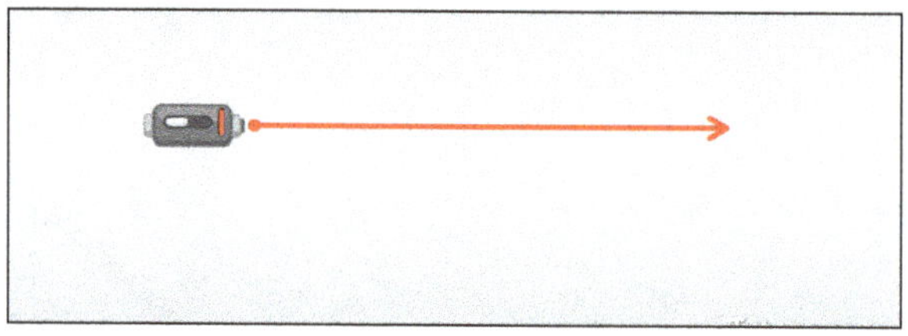

Line Segments

In geometry, a line segment is bounded by two distinct points on a line. Or we can say a line segment is part of the line that connects to points. A line has no endpoints and extends infinitely in both the direction but a line segment has two fixed or definite endpoints. Line segment and a ray have a difference that a ray has only one endpoint, which is its starting point.

Most of the shapes that we see around us are either made up of lines or curves. Various geometrical shapes are a combination of these. A one-dimensional collection of points extending infinitely in either direction is a line and any portion of a line which has two endpoints is a line segment.

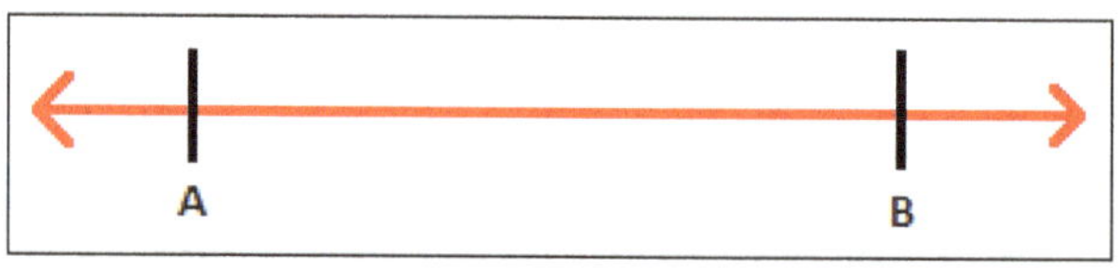

The above figure represents a line segment with two endpoints A and B and is denoted by the bar symbol (—) such as $\overline{AB}$. The length of a line segment can be varying and various methods are adopted to measure its length. To compare two or more line segments, a relationship between their lengths is established.

A line segment has two endpoints in a line. The length of the line segment is fixed, which the distance between two fixed points is. The length here can be measured by metric units such as centimeter (cm), millimeters (mm) or by conventional units like feet or inches. A closed line segment consists of both endpoints, whereas an open line segment is exclusive of the two endpoints. A line segment which has exactly one endpoint is called half-open line segment.

Measuring Line Segment

The most trivial method of comparison of two line segments is simple observation. Just by observing two line segments one can predict which is long or short compared to the other. In the figure below, by observation itself, we can say that the line segment CD is greater in length as compared to the line segment AB. But this method has several constraints, every time we cannot rely simply on observation to compare two line segments.

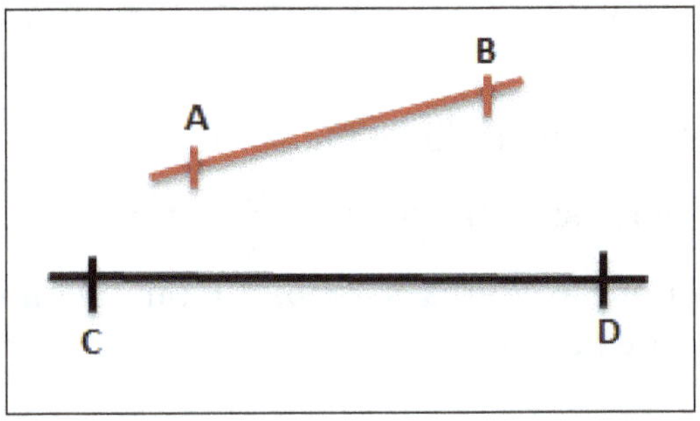

Using Trace Paper

With the help of trace paper, two line segments can be easily compared. Trace one line segment and place it over on the other segment and it can be easily deduced which is greater in length. For more than two line segments repeat the process again and again. For precise comparison, the line segments must be traced accurately. Hence this method depends on the accuracy of tracing which puts a limitation on this method.

Using Ruler and Divider

There are certain markings on the ruler beginning from zero as shown in the figure given below; these markings divide the ruler into equal parts. Each part is equal to a length of 1 cm and these unit centimeters are further subdivided into 10 parts and each sub-part is equivalent to 1 millimeter.

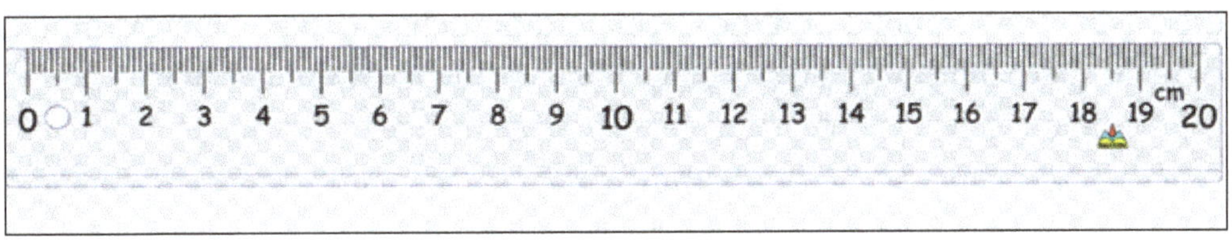

To measure a line segment AB, place the zero marking of the ruler along the beginning of the line and measure its length accordingly. In the figure given below, the length of line segment AB is 8 cm.

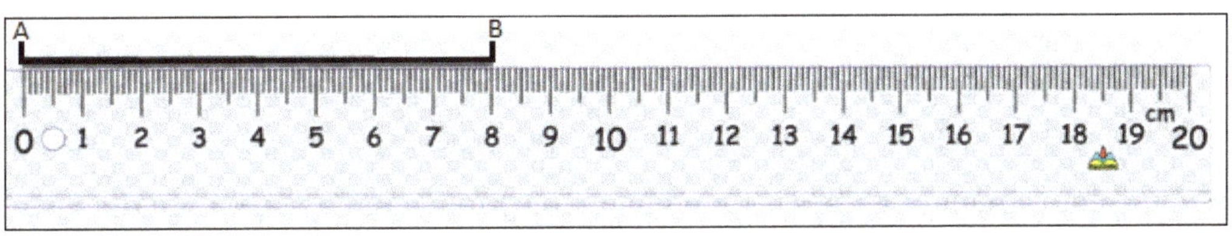

To eliminate the positioning error we use a divider. Place one of the needles of the divider at A and other at B and then place the divider along the ruler and measure its length. This method is more accurate and reliable.

Construction of Line Segment

Here we will learn to draw a line segment with the help of a compass and measuring ruler or scale. Let us assume we need to draw a line segment of length 7cm. Then follow the below given steps:

- Draw a line of any length without a measurement (keeping in consideration of the length of the line segment).
- Mark a point P on the line, which will be the starting point of the line segment.
- Now using a scale or ruler, locate the pointer of compass 7cm apart from the pencil's lead.
- Again, place the pointer of the compass at point P on the line and with the same measurement mark an arc with the pencil.
- Now, mark this point as Q.

Hence, PQ is the required line segment of length 7cm.

Angle

Angles are very useful in our daily life so it's important to learn and understand about them. For example, the minute hand of a wall clock turns angle 360 degrees to make a minute. It takes the earth 24 hours to rotate an angle of 360°. And that's how the clocks are designed to synchronize with the rotation of the Earth. For many such reasons, it is therefore necessary to study angles.

In mathematics, an angle is defined as a geometric figure created by two rays sharing a common end point. The rays or lines that join or intersect at a common point to form an angle are referred to as sides of the angle. The units for measuring angles are the degree (°) and radian (rad). Angles are represented using the symbol ∠ and the Greek letters such as θ, α etc. For example, ∠ ABC = θ. Here, points A and C are the sides of the angle while B is the vertex of the lines.

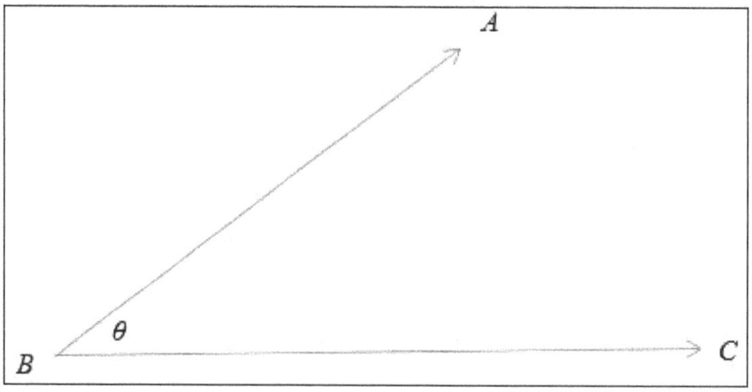

The concept of measurement of angles is dated back to 1500 BC in Egypt, where the Sun' s shadows were taken against graduations marked on stone tablets. The best evidence of this method can be seen in the Egyptian Museum found in Berlin. A shadow was cast over a graduated stone tablet using a vertical rod known as "Gnomon." With this method, Egyptians were able to accurately measure time and seasons.

Measurement of Angles

The modern way of measuring an angle is by the use of a protractor. A protractor is a transparent glass or plastic tool that has calibrations in either in radian or degree scale. You can read angles from a protractor either by moving clockwise or anticlockwise.

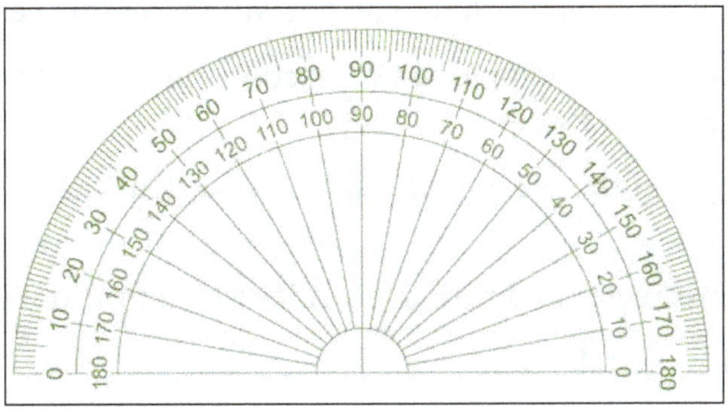

To measure an angle using a protractor, line up one line or ray along the zero-degree line on the protractor. Make sure the vertex of lines coincides with the midpoint of the protractor. Finally, follow the second line to read to the nearest degree the size of the angle.

Example: To measure angle "α = 300" between two lines, proceed as follows:

Line up one line along 0° mark on the protractor and follow the second line to read the angle.

∠ a = 30°

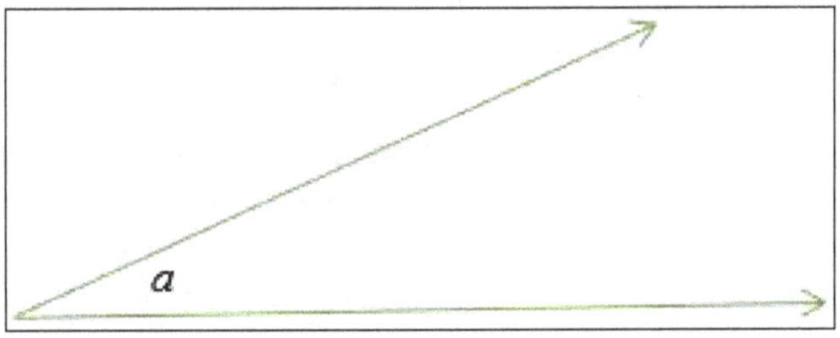

Example: Line up one line along 0° mark on the protractor and follow the second line to read the angle.

∠ β = 90°

How to Draw Angles?

Angles less than 180° can be drawn using a protractor by following the steps below:

- Draw a straight line of any measurement. The straight line will act as the arm of the angle.
- Mark a dot at any position on the line. The dot will represent the vertex of the angle.
- Place the middle of the protractor at the vertex dot and ensure the mark on the line and the center of the protractor flashes.
- Find the given angle on the protractor scale and mark a small dot at the edge of the protractor.
- Now, remove the protractor and then use a ruler to joint the dot with the vertex mark.
- Label the angle with capital letters.

Example: Draw ∠ ABC = 50°.

Explanation:

- Draw a straight line of any dimension and place a dot along the line.

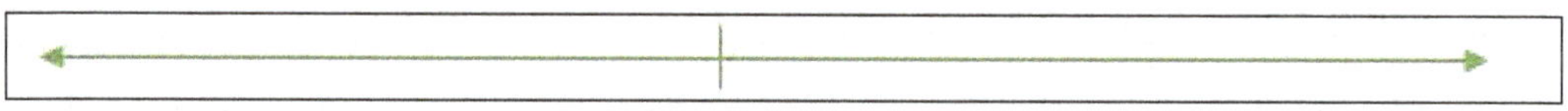

- Place the base of the protractor on the line such that its center coincides with the mark.
- Move anticlockwise along the protractor scale and place a dot at 50°.
- Using a ruler, join the dot with the mark on the line.
- Label you diagram such that A and C are the sides and B is the vertex of angle.

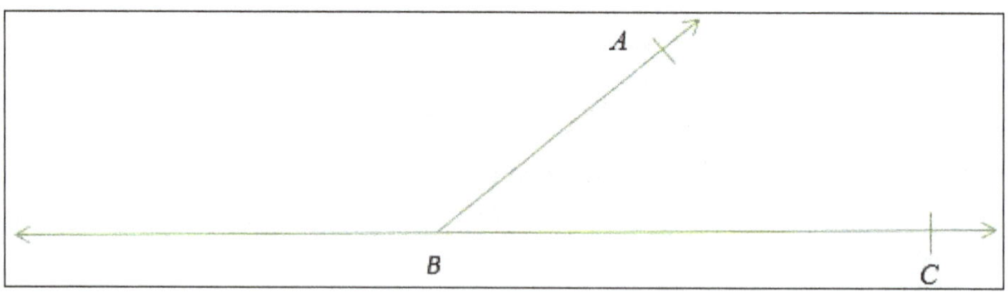

You can use the same procedure to draw any angle less than 180°.

Angles Pairs

Angle pair is angles that appear in twos to display a certain geometrical property. Some of the angle pairs include:

- Complementary angles,
- Supplementary angles,
- Vertical angles,

- Alternate interior angles and alternate exterior angles,
- Corresponding angles,
- Adjacent angles.

Complementary Angles

Two angles are said to be complementary to each other if their sum equals to 90° (right angle).

Explanation of complementary angles:

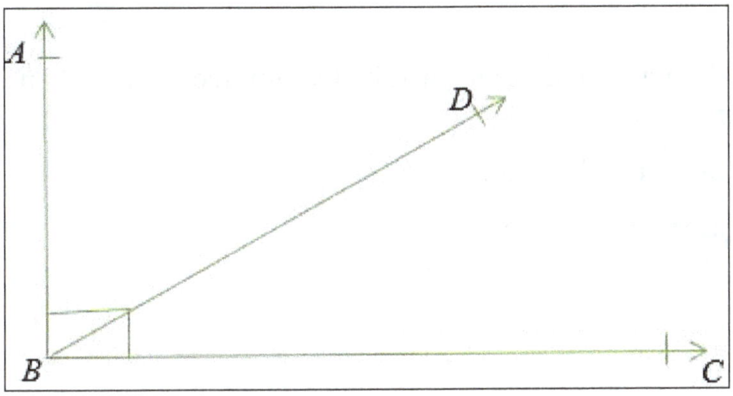

∠ ABD is a compliment of ∠ DBC because; ∠ ABD + ∠ DBC = 90° (right angle).

Examples of complementary angles:

- 40° and 50°
- 60° and 30°
- 85° and 5°
- 70° and 20°
- 45° and 45° etc.

Supplementary Angles

∠ ABD is a supplement of ∠ DBC because; Supplementary angles are pair angles whose sum of degree measurements equals to 180° (straight line).

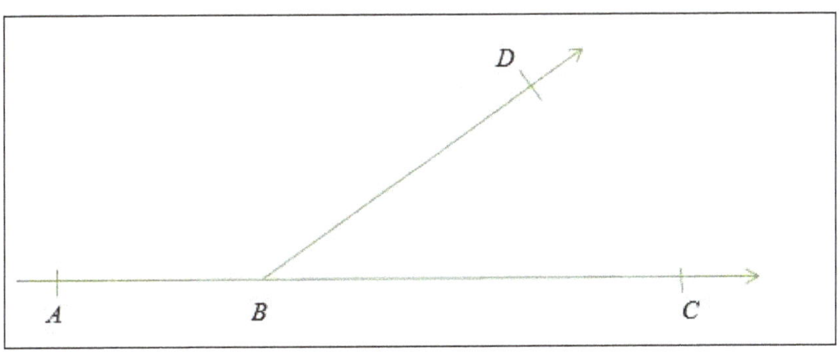

∠ ABD + ∠ DBC = 180° (straight line).

Some of the examples of supplementary angles include:

- 140° and 40°
- 120° and 60°
- 30° and 150°
- 90° and 90° etc.

Vertical Angles

Vertical angles are pair angles that are formed by two intersecting lines such that the angles are opposite to each other.

- ∠ a = ∠ d and are vertical angles.
- ∠ c = ∠ b and are vertical angles.

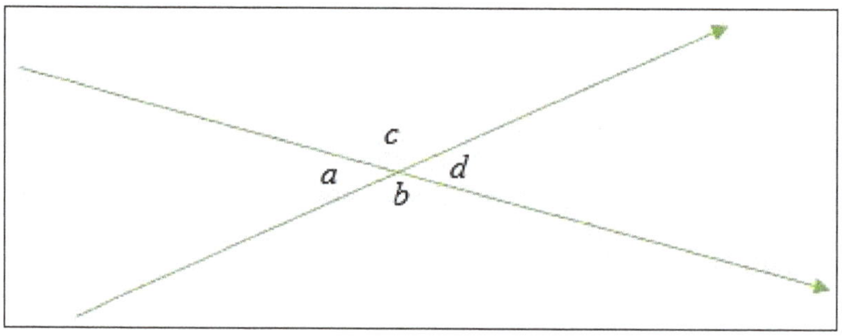

Alternate Interior Angles

Alternate interior angles are pair angles formed when a line intersect two parallel lines. Alternate interior angles are always equal to each other.

From the diagram below alternate angles are:

- ∠ c and ∠ e,
- ∠ b and ∠ f.

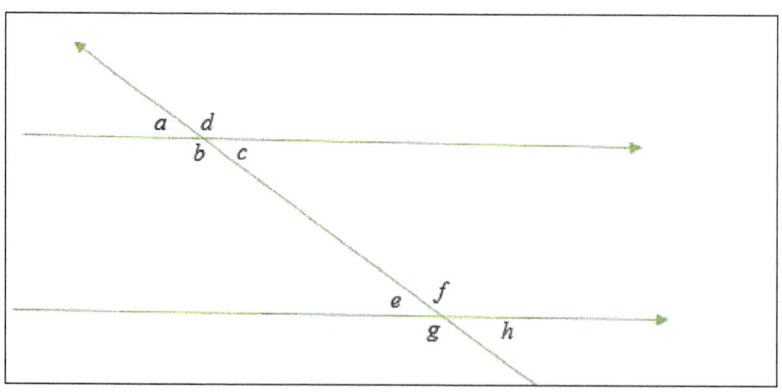

Alternate Exterior Angles

Alternate exterior angles are simply vertical angles of the alternate interior angles. Alternate exterior angles are equivalent.

Alternate exterior angles are:

- ∠ a and ∠ h,
- ∠ d and ∠ g.

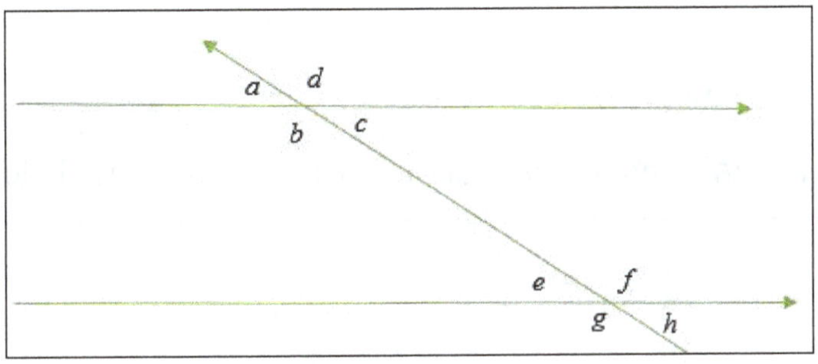

Corresponding Angles

Corresponding angles are pair angles formed when a line intersects a pair of parallel lines. Corresponding angles are also equal to each other.

Pairs of corresponding angles from the diagram below are:

- ∠ a and e,
- ∠ d and ∠ f,
- ∠ c and ∠ h,
- ∠ b and ∠g.

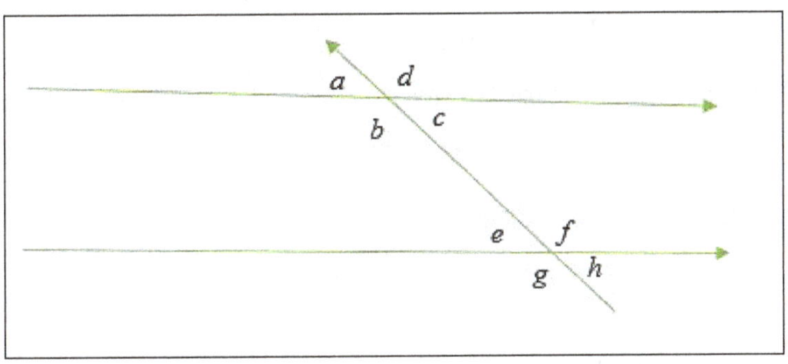

Adjacent Angles

Adjacent angles are pair angles that are next to one another. They share a common vertex and side.

Some of the example of adjacent angles from the figure below are:

- ∠ a and ∠ d,

- ∠ d and ∠ c,
- ∠ c and ∠ b etc.

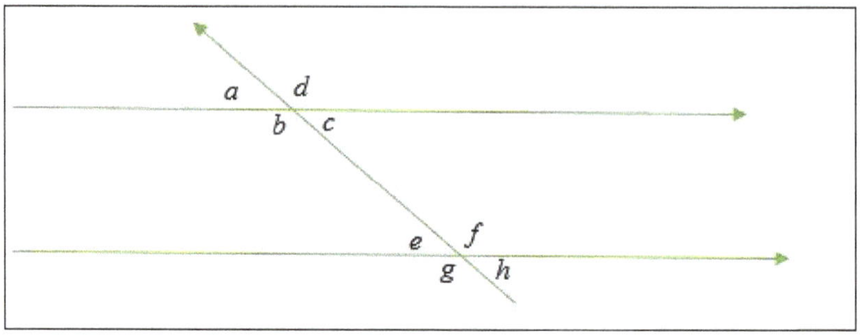

Example: Given that ∠ a = 45°, find all the other angles in the diagram below:

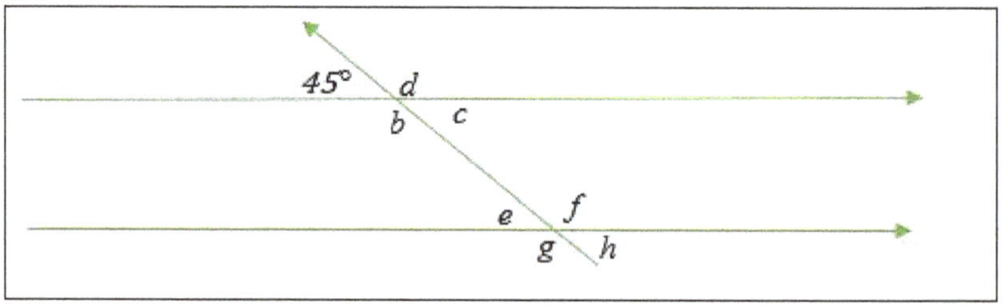

Solution: Given ∠ a = 45°.

- ∠ a and ∠ d are supplementary angles (add up to 180°). Therefore ∠ d = 180° – 45°.
- ∠ d = 135°.
- ∠ a and ∠ c are vertical angles, and vertical angles are equal. Therefore ∠ c = 45°.
- ∠ d = 135° and ∠ b are vertical angles.
- ∠ b = 135°.
- ∠ a = ∠ e = 45° (corresponding angles).
- ∠ d and ∠ f = 135° (corresponding angles).
- 2 b =∠ g= 135° (corresponding angles).
- ∠ c = ∠ h = 45° (corresponding angles).

Types of Angles

Different types of angles exist in nature and each one of them carries much importance in our daily lives. For example, architects and engineers use angles when designing machines, buildings, roads and bridges. In sports, athletes use angles to enhance their performance, for example, in short put, a person must spin with the disk a certain angle in order to throw it far. In soccer, you must use a certain angle in order to pass the ball to the next player.

Carpenters and artisans also use angles to fabricate objects such as sofas, tables, chairs, pails, and other. Artists use angles to sketch portraits and paintings. Fashion designers also use angles to come with certain with the best outfits. For these reasons, it is therefore necessary we learn the different types of angles.

Different Types of Angles

Angles are classified on the basis of:

- Magnitude,
- Rotation.

Classification of Angles based on their Magnitude

There are seven types of angles based on their degree measurement. They include:

- Zero Angles,
- Acute Angles,
- Right Angles,
- Obtuse Angles,
- Straight Angles,
- Reflex Angles,
- Complete Angle.

Zero Angle

A zero angle (0°) is an angle formed when the both the arms of the angle are at the same position.

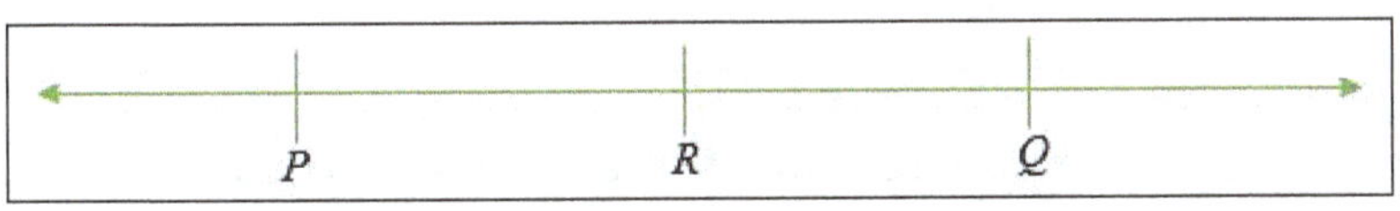

∠ RPQ = 0° (zero angle).

Acute Angle

An acute angle is an angle that is more than 0° but less than 90°. Common examples of acute angles include: 15°, 30°, 45°, 60° etc.

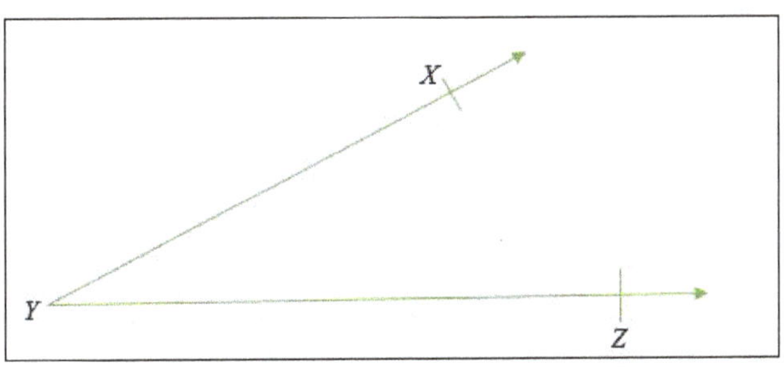

∠ XYZ is greater than 0° but less than 90° (acute angle).

90 Degree Angle

A 90-degree angle, also known as a right angle is an angle whose measure is equal to 90° is called a right angle. Right angles are represented by drawing a small square box between the arms of an angle.

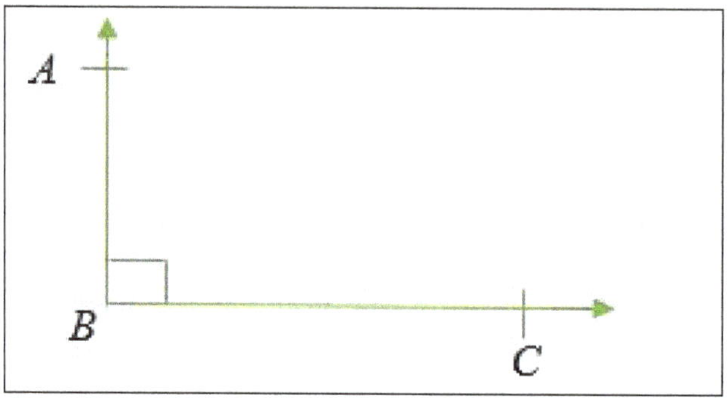

∠ ABC = 90° (right angle).

Obtuse Angle

An obtuse angle is a type of angle whose degree measurement is more than 90° but less than 180°. Examples of obtuse angles are: 100°, 120°, 140°, 160°, 170° etc.

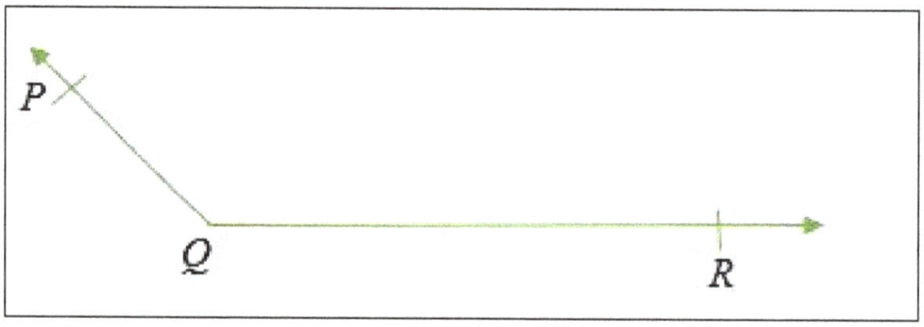

∠ PQR is an obtuse angle because it's less than 180° and greater than 90°.

Straight Angle

As the name suggests, a straight angle is an angle whose measure is equal to 180° (straight line).

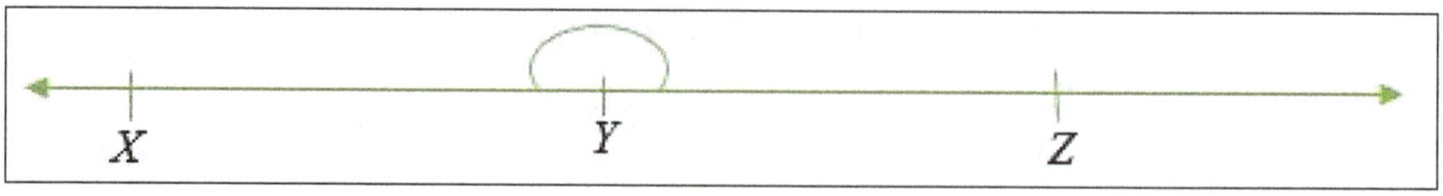

∠ XYZ = 180° (straight angle).

Reflex Angle

Reflex angles are the types of angles whose degree measurement is more than 180° but less than 360°. Common examples of reflex angles are: 200°, 220°, 250°, 300°, 350° etc.

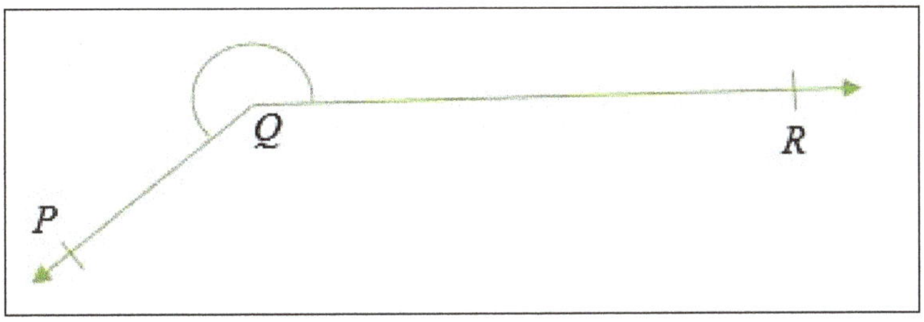

∠ PQR is greater than 180° but less than 360°.

Complete Angle

A complete angle is equal to 360°. 1 revolution is equal to 360°.

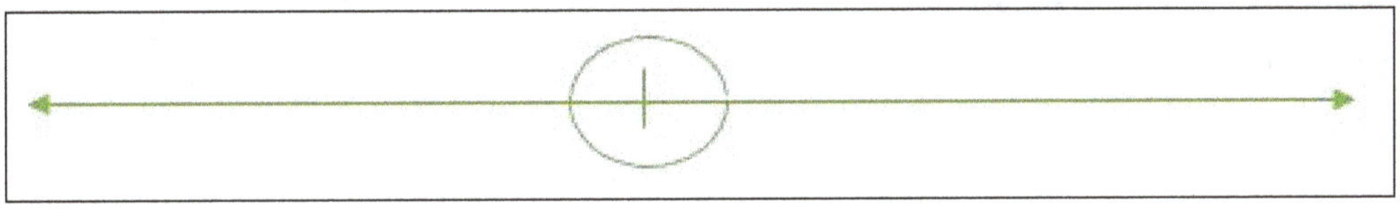

Classification of Angles on the basis of Rotation

Based on direction of rotation, angles can be classified into two categories, namely:

- Positive Angles: Positive angles are the types of angles whose measurement are taken in a counterclockwise direction from the base.
- Negative Angles: Negative angles are measured in a clockwise direction from the base.

Other Types of Angles

There are other types of angles known as pair angles. They are called pair angles because they appear in pairs to show a certain property. These are:

- Adjacent angles have the same vertex and arm.
- Complementary angles: Pair angles that add up to 90°.
- Supplementary angles: Pair angles whose sum of angles is equal to 180°.
- Vertically Opposite Angles: Vertically opposite angles are equal.
- Alternate Interior Angles: Alternate interior angles are pair angles formed when a line intersects two parallel lines. Alternate interior angles are always equal to each other.
- Alternate Exterior Angles: Alternate exterior angles are simply vertical angles of the alternate interior angles. Alternate exterior angles are equivalent.
- Corresponding Angles: Corresponding angles are pair angles formed when a line intersects a pair of parallel lines. Corresponding angles are also equal to each other.

Example: A certain angle is such that, two times the sum of its size and 10° is 60°. What is the name of this angle?

Solution: The angle be x°.

$$\Rightarrow 2(x+10)=60°$$
$$\Rightarrow 2x + 20°=60°$$
$$\Rightarrow 2x=40°$$
$$x=20°$$

The angle is 20°. Since 20° is less than 90°, the type of the angle acute angle.

Example 2: The sum of three angles is (x+1), (x -1) and (x + 3) forms a right angle. Find value of x.

Solution:

$$\Rightarrow (x+1)+(x-1)+(x+3)=90$$
$$\Rightarrow 3x+3=90$$
$$\Rightarrow 3x=87$$
$$x=29$$

Therefore, the value of x is 29 degrees.

Example: The sum of three angles (5x + 4), (x – 2) and (3x + 7) forms a straight angle. Find the type of these angles.

Solution:

$$\Rightarrow (5x+4)+(x-2)+(3x+7)=180°$$
$$\Rightarrow 5x+4+x-2+3x+7=180°$$
$$\Rightarrow 9x+9=180$$
$$\Rightarrow 9x=171$$
$$x=19$$

So, angle (5x + 4) = 99° ⇒ obtuse angle.

$$(x-2)=17° \Rightarrow \text{acute angle}$$
$$(3x+7)=64° \Rightarrow \text{acute angle}$$

Plane

A plane is an abstract concept that models the common notion of a flat surface — without depressions, protrusions, holes and a boundary — that for any two of its points entirely contains the straight line joining them.

Assuming a common (intuitive, physical) idea of the geometry of space, "plane" can be defined in terms of distances, orthogonality, lines, coordinates etc. In a more abstract approach (vector spaces) planes are defined as two-dimensional affine subspaces. In an axiomatic approach, basic concepts of elementary geometry, such as “point”, “line” and “plane”, are undefined primitives.

The surface of this metallic body consists of rectangles situated in various planes.

Non-axiomatic Approach

To define a plane is more complicated than it may seem. It is tempting to define a plane as a surface with zero curvature, where a surface is defined as a geometric object having length and breadth but no depth. However, this is not a good idea; such definitions are useless in mathematics, since they cannot be used when proving theorems. Planes are treated by elementary geometry, but the notions of surface and curvature are not elementary, they need more advanced mathematics and more sophisticated definitions. Fortunately, it is possible to define a plane via more elementary notions, and this way is preferred in mathematics.

The definitions of “plane” given below may be compared with the definition of a circle as consisting of those points in a plane that is a given distance (the radius) away from a given point (the center). A circle is a set of points chosen according to their relation to some given parameters (center and radius). Similarly, a plane is a set of points chosen according to their relation to some given objects (points, lines etc). However, a circle determines its center and radius uniquely; for a plane, the situation is different. Four equivalent definitions of “plane” are given below. Any other definition is equally acceptable provided that it is equivalent to these. Note that a part of a plane is not a plane. Likewise, a line segment is not a line.

Below, all points, lines and planes are situated in the space (assumed to be a three-dimensional Euclidean space), and by lines we mean straight lines:

- Definition via distances: Let two different points A and B be given. The set of all points C that are equally far from A and B — that is, | AC | = | BC | is a plane. This is the plane orthogonal to the line AB through the middle point of the line segment AB.
- Definition via right angles (orthogonality): Let two different points A and B be given. The set of all points C such that the lines AB and AC are orthogonal (that is, the angle BAC is right) is a plane. This is the plane orthogonal to the line AB through the point A.

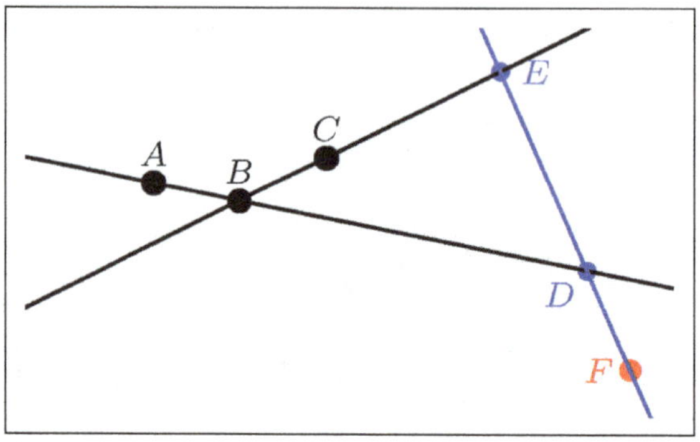

Every point F on the line DE belongs to the plane through given points A, B and C, provided that D belongs to the line AB, E belongs to the line BC, and D, E do not coincide with B.

- Definition via lines: Let three points A, B and C be given, not lying on a line. Consider the lines DE for all points D (different from B) on the line AB and all points E (also different from B) on the line BC. The union of all these lines, together with the point B, is a plane. This is the plane through A, B and C.
- In other words, this plane is the set of all points F such that either F coincides with B or there exists a line through F that intersects the lines AB and BC (in distinct points).
- Definition via Cartesian coordinates: In terms of Cartesian coordinates x, y, z ascribed to every point of the space, a plane is the set of points whose coordinates satisfy the linear equation $ax + by + cz = d$. Here real numbers a, b, c and d are parameters such that at least one of a, b, c does not vanish.

Some Properties of Planes

Most Basic Properties

- For any three points not situated in the same straight line there exists one and only one plane that contains these three points.
- If (at least) two points of a given straight line lie in a given plane, then all points of this line lie in that plane.
- If two planes have a common point then they have (at least) a second point in common.
- Every plane contains (at least) three points not lying in the same straight line.
- The space contains (at least) four points not lying in the same plane.
- Further properties.
- Two planes either does not intersect (are parallel), or intersect in a line, or coincide.
- A line either does not intersect a plane (is parallel to it), or intersects it in a single point, or is contained in the plane.
- Two lines perpendicular to the same plane are parallel to each other (or coincide).
- Two planes perpendicular to the same line are parallel to each other (or coincide).

Modern Approach

The modern approach defines the three-dimensional Euclidean space more algebraically, via linear spaces and quadratic forms, namely, as a real affine space whose difference space is a three-dimensional inner product space. In this approach a plane in an n-dimensional affine space (n = 3, or even more) is defined as a two-dimensional affine subspace.

A less formal version of this approach uses points, vectors and scalar product (called also dot product or inner product) of vectors without mentioning linear and affine spaces. Optionally, Cartesian coordinates of points and vectors are used. There, in particular, equivalence between the definition via right angles (orthogonality) and the definition via Cartesian coordinates is explained.

Plane Geometry

Plane geometry (also called "planar geometry") is a part of solid geometry that restricts itself to a single plane ("the plane") treated as a geometric universe. In other words, plane geometry is the theory of the two-dimensional Euclidean space, while solid geometry is the theory of the three-dimensional Euclidean space.

Plane geometry studies the properties of plane figures (and configurations). Plane figures in elementary geometry are sets of points, lines, line segments and sometimes curves that fall on the same plane. For example: Triangles, polygons and circles. In plane geometry every figure is plane, in contrast to solid geometry.

Algebraic Equations

In analytic geometry several closely related algebraic equations are known for a plane in three-dimensional Euclidean space.

Point-normal Representation

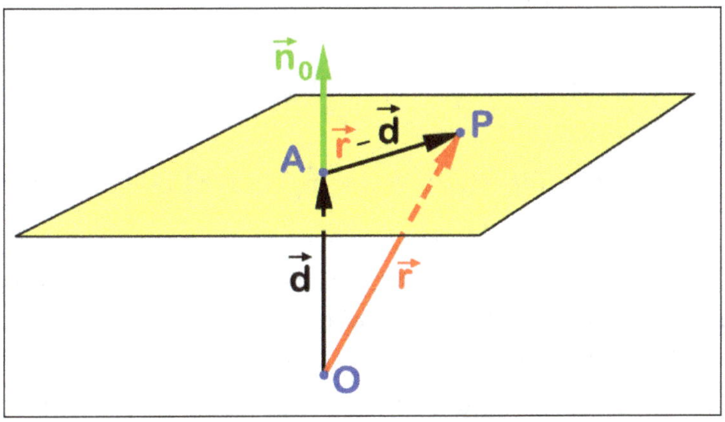

The equation for a plane: P is an arbitrary point in plane. In this picture A is chosen to be closest to the origin O, thus $\mathbf{d}$ and $\vec{\mathbf{n}}_0$ are collinear.

One such equation is illustrated in Figure above. Point P runs over the plane; point O is the origin (drawn outside the plane, but can also lie on the plane); A is a given point in the plane (in the picture it is the closest to the origin, but can be chosen arbitrarily); and vector $\mathbf{n}_0$ is a unit (length 1) vector normal (perpendicular) to the plane. Taking into account that two nonzero vectors are orthogonal if and only if their scalar product vanishes one gets the following relation (according to the definition via right angles):

$$(\mathbf{r}-\mathbf{d})\cdot\mathbf{n}_0 = 0$$

where,

$$\mathbf{d}=\overrightarrow{OA},\quad \mathbf{r}=\overrightarrow{OP},\quad \text{thus}\quad \mathbf{r}-\mathbf{d}=\overrightarrow{AP}.$$

This equation for the plane can be rewritten in terms of coordinates with respect to a Cartesian frame with origin in O:

$$(\mathbf{r}-\mathbf{d})\cdot\mathbf{n}_0 = 0 \Leftrightarrow \mathbf{n}_0\cdot\mathbf{r}=\mathbf{n}_0\cdot\mathbf{d} \Leftrightarrow ax+by+cz=d$$

where,

$$\mathbf{n}_0=(a,b,c),\quad \mathbf{r}=(x,y,z),\ \text{and}\ d=\mathbf{n}_0\cdot\mathbf{d}.$$

The definition via Cartesian coordinates is thus derived from the definition via right angles.

The parameters a,b,c,d of the equation do not depend on the choice of the point A on the plane, and satisfy $a^2 + b^2 + c^2 = 1$. It is convenient to choose A to be the closest to O point of the plane. Then the vector **d** is perpendicular to the plane and therefore collinear to $\mathbf{n}_0$; together with the relation $\mathbf{d}\cdot\mathbf{n}_0 = d$ it implies $\mathbf{d} = d\mathbf{n}_0$ and so, d is indeed the length of the vector **d** as well as the distance of the plane from the origin, provided that $d \geq 0$ (which can be ensured by the right orientation of $\mathbf{n}_0$.

Moreover, the Hesse normal form for the plane (called after the 19th century mathematician Ludwig Otto Hesse) is obtained,

$$\mathbf{r}\cdot\mathbf{n}_0 = d;$$

it is characterized by the use of a unit-length vector $\mathbf{n}_0$ rather than an arbitrary vector orthogonal to the plane.

Conversely, the definition via right angles can be derived from the definition via Cartesian coordinates as follows. Given a linear equation for a plane ax + by + cz = d where at least one of a, b, c does not vanish, we may restrict ourselves to the case $a^2 + b^2 + c^2 = 1$ (since the general case is reduced to this special case via dividing all terms by $\sqrt{a^2+b^2+c^2}$). Then we introduce,

$$\mathbf{r}=(x,y,z),\quad \mathbf{n}_0=(a,b,c),\quad \mathbf{d}=d\mathbf{n}_0$$

(or any other vector **d** satistying $\mathbf{d}\cdot\mathbf{n}_0 = d$), and get the same orthogonality relation,

$$(\mathbf{r}-\mathbf{d})\cdot\mathbf{n}_0 = 0.$$

Three-point Representation

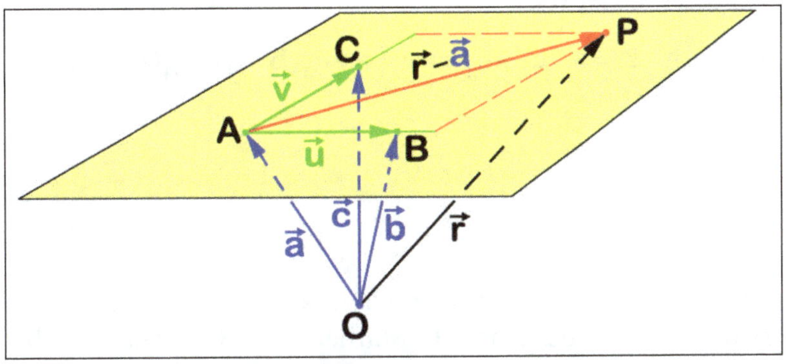

Plane through points A, B, and C.

Figure above shows a plane that by definition passes through three given points A, B, and C that are not on one line. The point P is an arbitrary point in the plane and the reference point O is again drawn outside the plane, but the case that the plane passes through O is not excluded. Referring to the figure we introduce the following definitions:

$$\mathbf{a} = \overrightarrow{OA}, \quad \mathbf{b} = \overrightarrow{OB}, \quad \mathbf{c} = \overrightarrow{OC}, \quad \mathbf{r} = \overrightarrow{OP}.$$

A plane, being a 2-dimensional affine space, becomes a 2-dimensional linear space if a chosen point A is treated as its origin. The following two vectors belong to the plane,

$$\mathbf{u} = \overrightarrow{AB} = \mathbf{b} - \mathbf{a}, \quad \mathbf{v} = \overrightarrow{AC} = \mathbf{c} - \mathbf{a}$$

and are non-collinear, therefore linearly independent. It follows that any vector in the plane is some linear combination of these two vectors. (This is also expressed as: Any vector in the plane can be decomposed into components along the two non-collinear vectors). In particular,

$$\overrightarrow{AP} = \mathbf{r} - \mathbf{a} = \lambda\mathbf{u} + \mu\mathbf{v} \quad \text{for some } \lambda, \mu \in \mathbb{R}.$$

The real numbers λ and μ specify the direction (and the length) of $\overrightarrow{AP}$. Hence the following equation for the position vector $\mathbf{r}$ of the arbitrary point P in the plane:

$$\mathbf{r} = \mathbf{a} + \lambda\mathbf{u} + \mu\mathbf{v}$$

is known as the point-direction representation of the plane. This representation is equal to the three-point representation,

$$\mathbf{r} = \mathbf{a} + \lambda(\mathbf{b} - \mathbf{a}) + \mu(\mathbf{c} - \mathbf{a}),$$

where $\mathbf{a}$, $\mathbf{b}$, and $\mathbf{c}$ are the position vectors of the three points that define the plane.

Writing for the position vector of the arbitrary point P in the plane,

$$\mathbf{r} = (1 - \lambda - \mu)\,\mathbf{a} + \lambda\mathbf{b} + \mu\mathbf{c} \equiv \xi_1\mathbf{a} + \xi_2\mathbf{b} + \xi_3\mathbf{c},$$

we find that the real triplet (ξ_1, ξ_2, ξ_3) with $\xi_1 + \xi_1 + \xi_1 = 1$ forms a set of coordinates for P. Such numbers ξ_1, ξ_2, ξ_3 are known as the barycentric coordinates of P. It is trivial to return from barycentric coordinates to the "three-point representation",

$$\mathbf{r} = \xi_1\mathbf{a} + \xi_2\mathbf{b} + \xi_3\mathbf{c} \quad \text{with} \quad \xi_1 = 1 - \xi_2 - \xi_3 \Leftrightarrow \mathbf{r} = \mathbf{a} + \xi_2(\mathbf{b} - \mathbf{a}) + \xi_3(\mathbf{c} - \mathbf{a}).$$

Curve

A curve is a shape or a line which is smoothly drawn in a plane having a bent or turns in it. For example, a circle is an example of curved-shape. In Mathematics, Geometry is a branch that deals with shapes, sizes, and the properties of figures. Geometry can be classified into two types. They are:

- Plane or Two Dimensional Geometry,
- Solid or Three Dimensional Geometry.

Two-dimensional geometry is the study of flat shapes like curves, lines, polygons, etc. that can be drawn on a piece of paper. On the other hand, three-dimensional geometry is the study of three-dimensional objects such as cubes, cylinders, spheres, and so on.

Curve Shape

Circles, ellipses, parabolas, and hyperbolas, as well as arcs, sectors and segments, are two-dimensional curved shapes. Whereas, shapes such as spheres, cylinders and cones, are referred to as three-dimensional curved shapes.

Types of Curves

Curves are classified into different categories. Let us have a look at different types of curves with examples:

Simple Curve

A curve that changes its direction, but it does not intersect itself. This type of curve is known as a simple curve. A simple curve may be open or closed.

Open simple curve Closed simple curve

Non-Simple Curve

The non-simple curve is a type of curve that crosses its path. It means the curve intersects itself while changing its direction.

Not simple closed curve Not simple open curve

Open Curve

A curve has two endpoints, and when it does not enclose the area within itself it is known as an open curve. Some of the examples of open curves are as follows:

Open curves

Closed Curve

A curve has no endpoints, and when it encloses the region or area will form, it is known as the closed curve. The type of curve is formed by joining the two endpoints of the open curve. The best example of closed curves are circles, ellipses, etc.

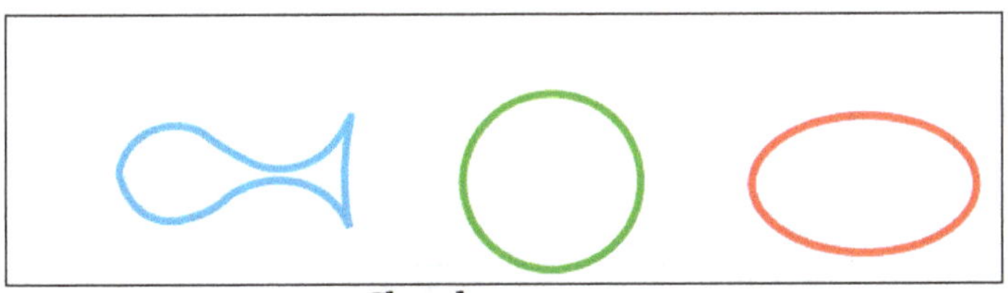

Closed curves

Upward Curve

A curve that points towards the upward direction is called an upward curve. The upward curves are called concave upward or convex downward curves.

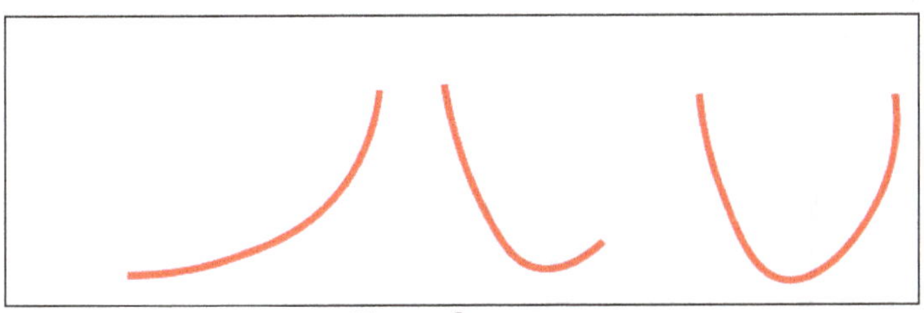

Upward curves

Downward Curve

A curve that points towards the downward direction is called a downward curve. The downward curves are called concave downward or convex upward curves.

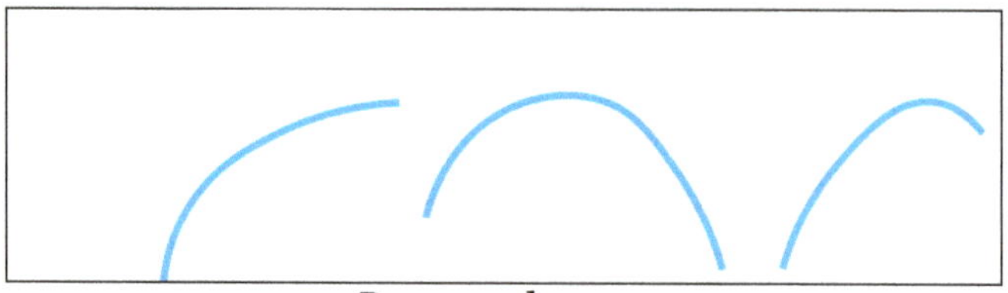

Downward curves

Curved Line

A curved line is a line that is not straight and is bent. If the curvature is not zero, then it is considered as a curve line. Generally, it is smooth and continuous.

Curved Line Images

We can observe many objects and things which are in the shape of a curved line. Some of them include railway track at the turning points, track of a roller coaster, paths of a road particularly for hill areas and so on. Apart from the real life examples, we can also observe the curve shaped lines in maths such as graph a quadratic polynomial, i.e. parabola, give curve, arrows, etc. Below are the images of curve shaped lines.

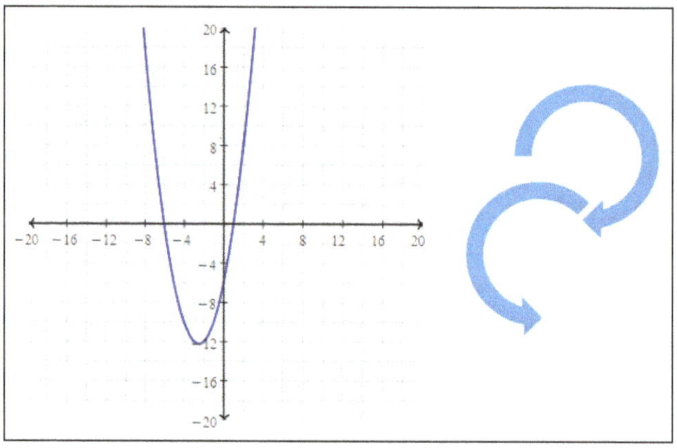

Area between the Curves

Area is a quantity that expresses the region covered for the given interval in the two-dimensional surface or a planar lamina. To find the area between the given two curves, it uses the concept of integration. The area between the two curves can be determined by calculating the difference between the integrals of two functions.

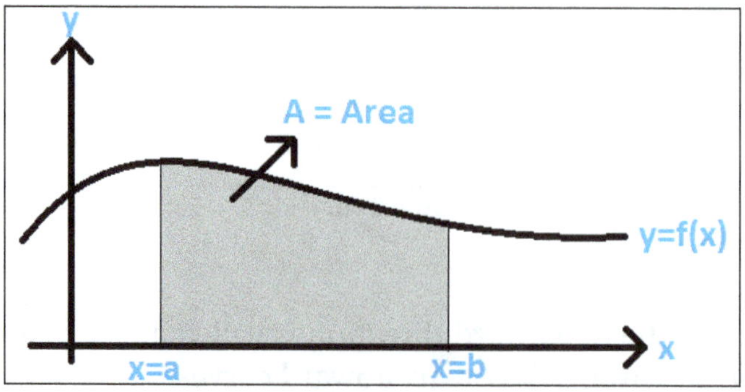

The area between two curves that is measured between two values "a" and "b" is given by the integral from a to b of the given function. Here, the value of b is defined as the larger value than the value of a. Therefore, the area between the two curves is defined as,

$$A = \int_a^b (f(x) - g(x))dx$$

where,

f(x) is the curve with greater value.

Manifold

A geometric object which locally has the structure (topological, smooth, homological, etc.) of $\mathbf{R}^n$ or some other vector space. This fundamental idea in mathematics refines and generalizes, to an arbitrary dimension, the notions of a line and a surface. The introduction of this idea was influenced by various requirements, both of mathematics itself and of other sciences. In mathematics, manifolds arose first of all assets of solutions of non-degenerate systems of equations and also as various sets of geometric and other objects allowing local parametrization; for example, the set of planes of dimension k in $\mathbf{R}^n$. They also appear as solutions of multi-dimensional variational problems (soap films), as integral manifolds of Pfaffian systems and of dynamical systems, as groups of geometric transformations and their homogeneous spaces, etc. In physics they play the role of models of space-time, in mechanics they act as phase spaces, energy levels, etc., in economics as surfaces of indifference, in psychology as spaces of sensations (for example, colours), etc.

Although the initial idea underlying the definition of a manifold is that of a local structure ("the very same as Rn"), this idea admits a whole series of global features typical for manifolds: (non-) orientability, homological Poincaré duality, the possibility of defining the degree of a mapping of one manifold onto another of the same dimension, etc. Of particular significance is the introduction of the tangent bundle and its related invariants.

The local structure of a manifold also allows the use of geometric techniques: putting into general position, the construction of Morse functions, etc., which are used for the geometric study of the global structure of manifolds, and, roughly speaking, consist of constructing a possibly simpler image of a manifold as a union of simple pieces; for instance, simplices or handles.

To make use of the idea of a manifold a transition from the local to the global point of view is usually made. The first stage is the introduction of a parametrization, that is, a representation of the "state space" of a given problem as a domain in a space of numbers $\mathbf{R}^n$. This gives one the possibility to describe by a set of numbers the coordinates of the corresponding point (the coordinate method). In the large the state space need not admit a similar description, that is, it need not be homeomorphic to a region in $\mathbf{R}^n$. If one does not resort to parametrizations with degeneracies (as in polar coordinates and their generalizations), then two paths are possible: either to introduce from the outset a larger than necessary number of parameters, and to find then the actual space implicitly using a system of equations ("state equations"), or to parametrize the space locally ("in the small"). For example, the set of lines in a plane is covered by two subsets: Π_1, consisting of lines with equations of the form y = kx + b, and Π_2, consisting of lines with equations of the form $x = \bar{k}y + \bar{b}$; both of these are homeomorphic to $\mathbf{R}^2$ with parametrizations by the pairs (k, b) and $(\bar{k}, \bar{b})$, respectively. However, in the large, this set is homeomorphic to an open Möbius strip, and hence not like $\mathbf{R}^2$.

When manifolds appear naturally in some domain or other, they are sure to carry some additional structure, which also is a topic of study in the domain. However, a major role is played by the topological structure, restricting the a priori possibilities. Conversely, in topology, the local and global properties of manifolds are studied taking into consideration the extra structures (for example, smooth) as tools.

The basis of the general idea of a manifold is the definition of a topological manifold as a topological space in which each point has a neighbourhood $\mathcal{X}$ and a homeomorphism $\phi : \mathcal{X} \to U$ onto a domain in $\mathbf{R}^n$ or into the half-space $\mathbf{R}^n_+ = \{x \in R^n : x_n \geq 0\}$; the homeomorphism φ is called a local parametrization, or chart, in $\mathcal{X}$. For a connected manifold M, the dimension n = dim M is an invariant. For a disconnected manifold the components are usually taken to be of the same dimension. A manifold decomposes into an interior Int M and a boundary ∂M(also called edge): points of the boundary correspond in the charts to the points of the boundary of $\mathbf{R}^n_+$, in $\mathbf{R}^n$. The boundary is an (n – 1)- dimensional manifold without boundary and need not be empty. A connected manifold without boundary is called open if it is non-compact, and closed if it is compact. The simplest examples of the four possible types of manifold are $\mathbf{R}^n, \mathbf{R}^n_+$, the ball B^n, and its boundary S^{n-1}. Although non-Hausdorff manifolds occur in certain situations (for example, the total space of a sheaf), it is usually assumed that a manifold is Hausdorff, paracompact, has a countable base, and, in particular, is metrizable.

The global specification of a manifold is accomplished by an atlas: A set of charts covering the manifold. To use manifolds in mathematical analysis it is necessary that the coordinate transitions from one chart to another are differentiable. Therefore differentiable are most often considered. A more general form is introduced by the idea of a Γ- structure on a manifold, given by atlases $\{\phi_i : \mathcal{X}_i \to U_i\}$ in which the coordinate transitions $h_{ij} = \phi_j \phi_i^{-1}$ are homeomorphisms in a system Γ of mappings of domains in $\mathbf{R}^n_+$ that is closed relative to composition. If Γ consists of r times continuously-differentiable mappings, then one says that the smoothness class of the manifold is Cr. In a similar way one defines analytic, and piecewise-linear, Lipschitz, etc., types of manifolds. Two Γ- atlases give a single Γ- structure if their union is a Γ- atlas. The classification of Γ- structures is the most important problem in the geometry of manifolds. A mapping f : M → N of one Γ- manifold into another is called a Γ- mapping if locally it has a "coordinate representation" $f_{ij} = \phi_j f \phi_i^{-1}$ where ϕ_j, ϕ_i are charts in M and N and $f_{ij} \in \Gamma$. In particular, there is the notion of a Γ- homeomorphism (a C^r- diffeomorphism if Γ = C^r).

Since, in mathematical analysis, manifolds are important as carriers of differentiable mappings, they are sometimes defined by their set of smooth functions, defined in neighbourhoods of points. The development of this idea has led to the idea of a pre-manifold or a ringed space (a sheaf of rings over a space), and further to the notion of a scheme. Replacing $\mathbf{R}^n$ by vector or other spaces, one arrives at various generalizations of a manifold such as, for example, complex-analytic manifolds. Infinite-dimensional manifolds arose in mathematical analysis and topology as spaces of mappings and sections of bundles, as spaces of homeomorphisms, spaces of closed subsets, etc. Their local models are vector spaces (Banach spaces, etc.) and spaces such as the Hilbert cube. The notions of smooth and other structures on infinite-dimensional spaces have not been sufficiently studied. A difficulty arises here because of the absence of technical theorems on approximations, existence of partitions of unity (a stock of small smooth functions), implicit functions, etc.

Manifolds being implicitly given as sets of solutions of systems of equations (inequalities in the case of a non-empty boundary) arise as subsets of $\mathbf{R}^n$. These manifolds are given all at once, and not in pieces, as with the assignment of an atlas. However, a necessary condition is non-degeneracy, otherwise every closed set can be given by one equation. The existence of a local parametrization is provided by the implicit-function theorem from the condition of maximal rank of the Jacobi matrix of the system. The equations are used as a language for expressing, by mathematical analysis, the properties of a manifold and are used to define the manifold. For example, the property of orthogonality of $(n \times n)$- matrices is given by a system of $n(n + 1)/2$ equations relative to the entries of the matrix. The system turns out to be non-degenerate, and the group of orthogonal matrices is a smooth submanifold in $\mathbf{R}^m$.

In a mathematical system with given coordinates, smaller systems are chosen by equations or inequalities expressing restrictions or "constraints". If the non-degeneracy condition for a system $\mathbf{F}_i = 0$ is satisfied at all points of a manifold, then the gradients of the functions $\mathbf{F}_i$ form a framing or frame (a k-frame orthogonal to the tangent plane at a point of the manifold and depending continuously on the point). Manifolds admitting a framing form the rather narrow class of stably-parallelizable manifolds. But, locally, any differentiable manifold in $\mathbf{R}^n$ may be given by a non-degenerate system, and by using a partition of unity it is possible to construct a system of constant (but not maximal) rank which defines the manifold.

For a manifold given by an atlas, the problem arises of realizing it as a submanifold in $\mathbf{R}^n$ with regard to some Γ- structure. Any topological, smooth or piecewise-linear manifold M is imbeddable, that is, is Γ-homeomorphic to a submanifold in $\mathbf{R}^{2n}$, and in $\mathbf{R}^{2n+1}$ the set of imbeddings is dense in the space of all continuous mappings. For other classes the question is significantly more complicated. It has been studied intensively, for example, for Riemannian manifolds. Algebraic, realized in a complex projective space (replacing here $\mathbf{R}^n$), constitute a very special. If degenerate systems of equations are allowed, then manifolds with singularities arises. In algebraic geometry both ideas combine: An algebraic space is defined by joining pieces which are locally given by polynomial systems with degeneracies. In cobordism theory, in topology, under the name of manifold with singularities one discusses spaces which are constructed locally as products of cones over manifolds. Subsequent generalizations are connected with the consideration of continuous families of manifolds.

The study of manifolds by pure topological means is very complicated and, until recently, was restricted to local properties, that is, virtually properties of $\mathbf{R}^n$ [Continuous decomposition|continuous decomposition]; wild imbedding; etc. The global study of manifolds necessarily requires the use of either homological or other means from algebraic topology, or smooth, piecewise-linear or other structures. The fundamental homological property of a manifold is Poincaré duality, which is also preserved for homology manifolds, and is used for generalizations of the notion of a manifold, such as Poincaré complexes. The degree of a mapping is homological in nature; this is a basic idea which is also defined for pseudo-manifolds, i.e. spaces (usually complexes) which have open connected dense subsets that are manifolds.

The geometric study of manifolds, based on the use of the various structures, comes first of all from the local equivalence of manifolds with $\mathbf{R}^n$, and from the possibility, by approximating a mapping by smooth or piecewise-linear mappings, of putting them into general position, etc. The first problem here is the representation of a manifold as a space made up of simple pieces. The initial idea was that of triangulation, developing into the general idea of a complex. The difficult problems of triangulability and equivalence of triangulations were cleared up in the 1960's and 1970's. A more flexible tool has turned out to be handle decomposition of a manifold, equivalent to considering Morse functions. The basic theorems (on h- cobordism, the generalized Poincaré conjecture, etc.) were proved using the technique of successive simplification of this decomposition. This technique also acts as a geometric foundation for the classification theorems on smooth and piecewise-linear structures in a given homotopy type of manifolds. These theorems, however, required the use of subtle invariants, connected with the tangent bundle of the manifold. Of great significance was also the discovery of new homology theories: K-theory and the theory of cobordism. Two manifolds are cobordant if together they bound a third manifold. Here the various structures are taken into account. This equivalence relation is based on the initial definition of homology, when cycles were considered naively as piecewise-smooth manifolds. In cobordism theory this idea was restored, but submanifolds were replaced by mappings of manifolds. The fundamental groups play a special role in questions on the classification of structures on manifolds, due to which the connection of these questions with algebraic K- theory has been established.

The most important invariant of a smooth manifold M is its tangent bundle τM. The subject of analysis on manifolds is the study of sections of τM and the various bundles associated with it. From the topological point of view τM is characterized by the existence of the exponential mapping $\exp : \tau M \to M$, diffeomorphically mapping small balls in the fibres $\tau_x M$ onto neighbourhoods of the corresponding points $x \in M$. It can be described as a projection of a tubular neighbourhood of the diagonal in $M \times M$ onto a factor. In this form the definition of τM can be transferred to the topological, piecewise-linear, etc., cases.

Different reductions of the structure group of the tangent bundle are called structures on the manifold. If an atlas can be chosen such that the Jacobi matrices of its coordinate transitions belong to a reduced structure group, and thus determine this reduction, then such a structure is called integrable. The question of classification of structures and their integrability is related to the fundamental questions of the differential geometry of manifolds. In the topology of manifolds the question on the classification of smooth, piecewise-linear and other structures reduces mainly to the classification of the corresponding structures on the tangent bundle and, by the same token, lead to homology problems.

The major role of the tangent bundle is connected with the fact that there is cohomology classes invariantly connected with it in some homology theory, carrying the most essential information on the global structure of the manifold and connecting the properties of the different constructions and structures on M with its topological properties.

Length

Length is the term used for identifying the size of an object or distance from one point to Length is a measure of how long an object is or the distance between two points. It is used for identifying the size of an object or distance from one point to another. The length of an object is its extended dimension, that is, its longest side. For example, the length of the ruler in the picture is 15 cm. The arrow above the ruler denotes the length of the ruler as it is the longest side of the ruler.

Some more examples:

The given image is of a rectangle, and for every rectangle, the length is longer than its breadth.

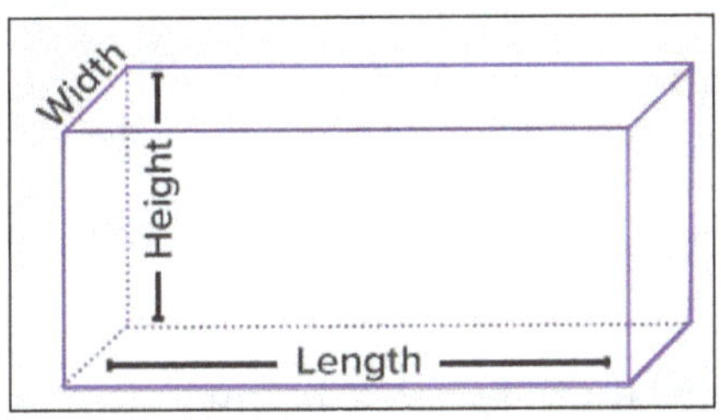

Tools for Measuring the Length

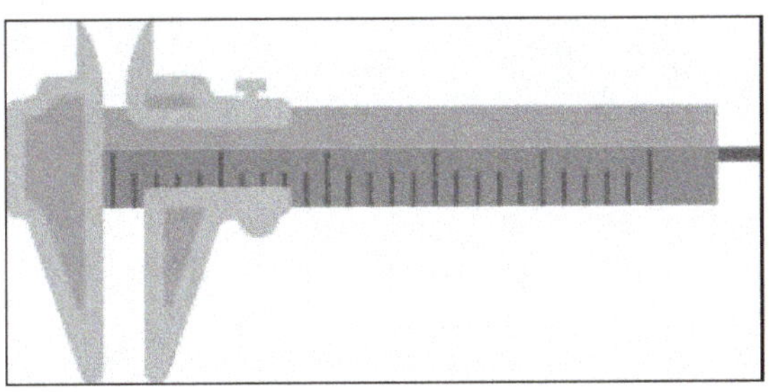

Vernier caliper

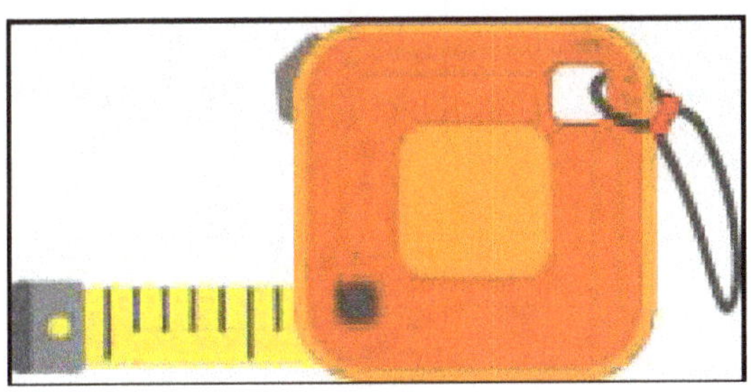

Tapes

Different Units of Length

The standard unit of length based on the metric system is a meter (m). According to the length that needs to be measured, we can convert a meter into various units like millimeters (mm), centimeter (cm), and kilometer (km).

Centimeters and millimeters help measure smaller lengths and meters and kilometers help measure larger lengths like distance. For example, the length of the pencils can be calculated in centimeters (cm), while kilometers can measure the distance between two buildings or places.

One hundred equal divisions of a meter give a centimeter. It is written as 'cm'. That is,

1 m = 100 cm

One thousand equal divisions of kilometer give a meter. That is,

1 km = 1,000 m

According to the length conversion charts, the different units of lengths and their equivalents are given below:

A kilometer (km), meter (m), and centimeter (cm) are the commonly used units of length. Conversion of these units is done using the given formula.

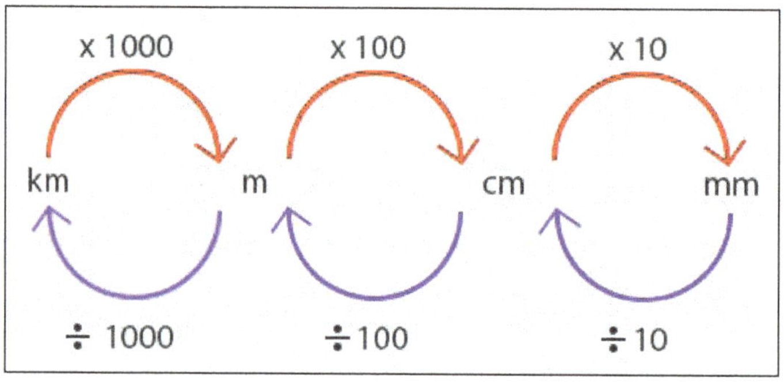

Additionally, in the customary system (followed in the United States) inches, feet, yards, and miles are used as the unit for length.

The relation between the customary units is given below:

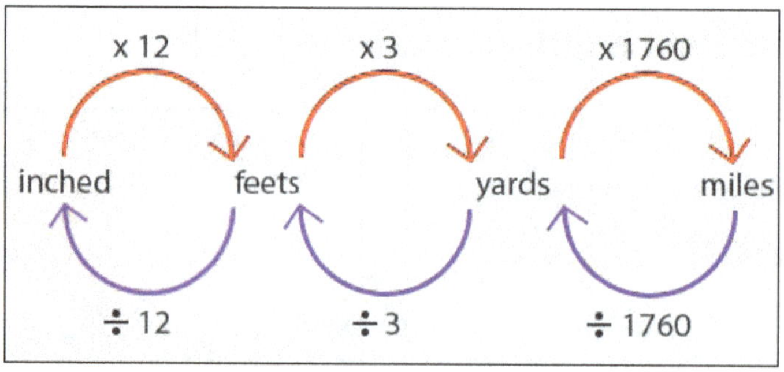

Metric System and Customary System

With the assortment of various units, the metric system seems quite a logical system as compared to the customary system and converting units in the metric system is much simpler than converting them in the customary system. The United States is the last remaining nation, which is still to adopt the metric system. However, it is easy to convert units in metric to the customary system by using the given conversion.

$1 \text{ meter } (m) = 39.4 \text{ inches} = 1.09 \text{ yards};$

$1 \text{ yard} = 0.92 \text{ m}$

$1 \text{ centimeter } (cm) = 0.39 \text{ inches};$

$1 \text{ inch} = 2.54 \text{ cm}$

Area

In geometry, area is the amount of space a flat shape - figures like a polygon, circle or ellipse takes up on a plane. Area confuses a lot of people because the area is measured in square units regardless of shape. Where is the number of square units in a circle? Once you know how square units relate to area, you can find the area of just about any two-dimensional shape.

How to Find the Area of a Shape

Flat shapes have two dimensions:

- Width
- Length

A square, for instance, has a width equal to its length because all side lengths are the same. An ellipse has width and length, too. We can easily see how the square could be divided up into small, square units like on a coordinate plane. You cannot easily see how an ellipse could be made up of little squares, but it can be. Since it has width and length, it covers a space, and that space, even with the curving sides of the ellipse, can be divided up into square units:

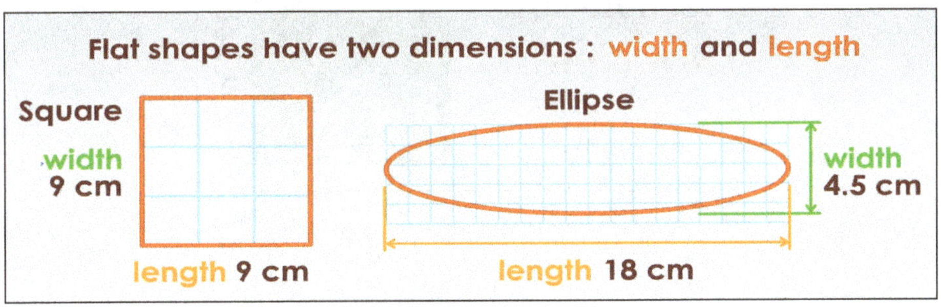

Counting the square units in the square is easy: one, two, three, etc. But, how can you count all the square units in the ellipse? How do you decide what part of a square is under the top curve? What about the curves at the left and right ends? Fortunately, mathematics has a fast way to add up all the square units without actually counting them.

Square units are the measurement unit for area because plane figures or flat shapes can always be divided into squares of known dimensions, like these:

- mm^2
- cm^2

- ft^2
- yd^2
- km^2
- mi^2

Whether you are finding the area of a quadrilateral like a trapezoid and a rhombus, or any other closed figure, the area will always be squared.

Area Formula

The area formula you use depends on which shape you are trying to find the area for.

Area of Squares and Rectangles

To find the area of simple shapes like a square or the area of a rectangle, you only need its width, w, and length, l (or base, b). The area is length time's width:

$$A = l \times w$$

The area is always squared. You will always express area as square units, derived from the linear units. Here is a rectangle 90 meters wide and 120 meters long (the largest size of a FIFA soccer field). What is its area of this rectangle?

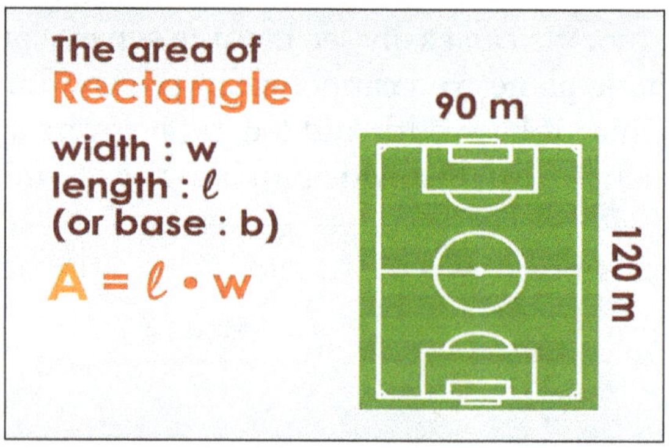

$$A = l \times w$$

$$A = 120m \times 90m$$

$$A = 10{,}800\,m^2$$

Because the soccer field is measured in linear meters, its area is square meters. The area of the rectangle is 10,800 meters squared.

The area of a square formula is actually even easier than writing out length × width because all sides are equal:

$$A = s^2$$

Here is a square with sides 15 inches long, the same size as the bases on an MLB baseball field. Calculating area for this square looks like this:

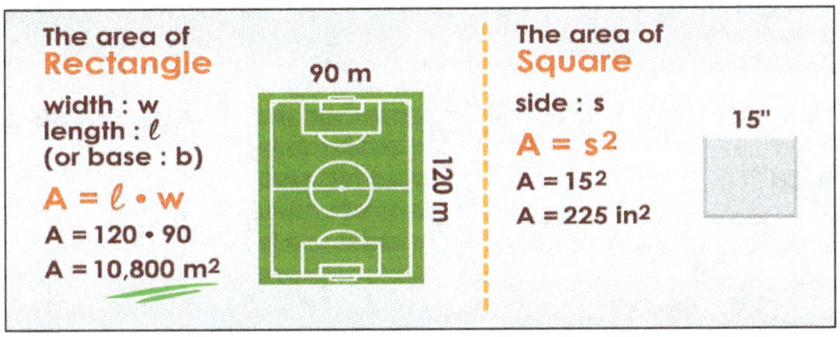

$A = s^2$

$A = 15^2$

$A = 225 \text{ in}^2$

Area of other Shapes

All the other polygons do not easily divide into square units. Take a look at a parallelogram:

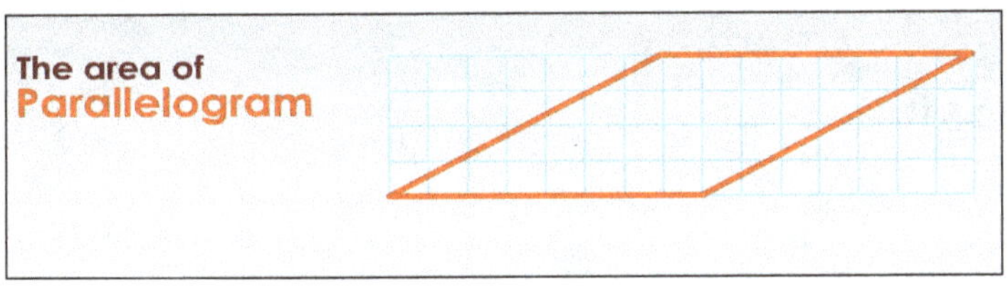

The two sides cut right across many square units. Of course, a parallelogram is just a knocked-over rectangle. So, mathematically, if we could cut off one end and attach it to the other, we would have the area in square units. We can do exactly that, since the area of a parallelogram with a base, b, and width or height, h, is found using this formula:

$A = b \times h$

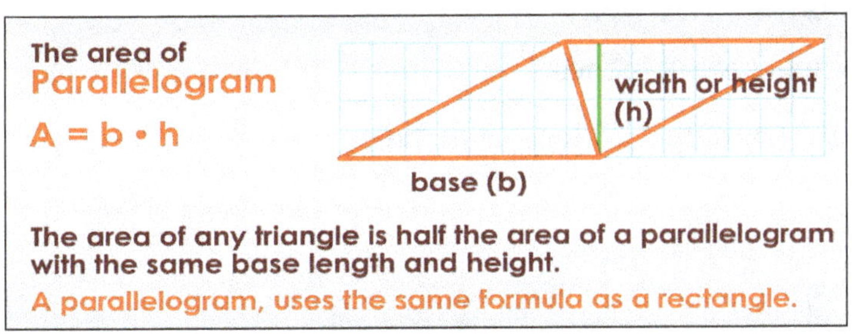

That is the same formula as for a square or rectangle.

If you divide a parallelogram along a diagonal, what do you have? Two triangles. That means the area of any triangle is half the area of a parallelogram with the same base length and height. A parallelogram, remember, uses the same formula as a rectangle.

The area of a triangle is, then, half the base, b, multiplied times the height, h:

$$A = \frac{1}{2}bh$$

Here is a right triangle, a sail from a 45-foot Morgan sailboat with a base $20\frac{1}{4}$ feet and a height $44\frac{1}{2}$ feet. What is its area?

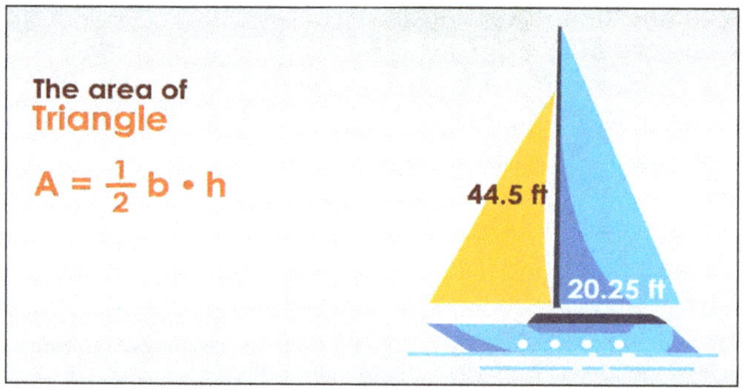

For convenience in multiplying, you can change the fractions to decimals:

$$A = \frac{1}{2}bh$$

$$A = \frac{1}{2}(20.25\text{ ft.} \times 44.5\text{ ft.})$$

$$A = \frac{1}{2}\left(901.125\text{ ft}^2\right)$$

$$A = 450.5625\text{ ft}^2$$

The area of the triangle sail is approximately 450.6 square feet.

How about the home plate of an MLB baseball field? We can calculate the area of the home-plate pentagon by considering it as two shapes:

- A rectangle 17 inches x 8.5 inches.
- An isosceles right triangle with legs 12 inches.

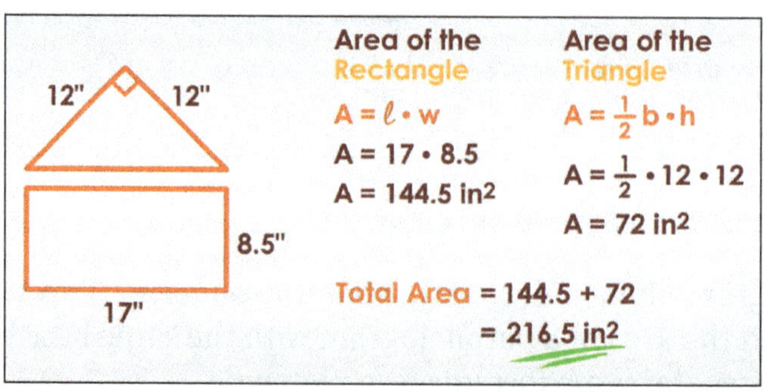

First, we'll use the formula to find the area of the rectangle:

$$A = l \times w$$

$$A = 17 \text{ inches} \times 8.5 \text{ inches}$$

$$A = 144.5 \text{ in}^2$$

Now, we will calculate the area of the remaining triangles using the area of a triangle formula:

$$A = \frac{1}{2}bh$$

$$A = \frac{1}{2}(12 \text{ inches} \times 12 \text{ inches})$$

$$A = \frac{1}{2}(144.5 \text{ in}^2)$$

$$A = 72 \text{ in}^2$$

Add these two areas to find the total area in square inches:

$$144.5 \text{ in}^2 + 72 \text{ in}^2 = 216.5 \text{ in}^2$$

Find the Area of a Circle

Some two-dimensional shapes are not even polygons, like our ellipse, or a circle. The area of a circle with radius r is found using this formula:

$$A = \pi r^2$$

If you have a circle with a radius of 4 centimeters, you can calculate the area of the circle easily with the formula above:

$$A = \pi r^2$$

$$A = \pi(4)^2$$

$$A = \pi(16)$$

$$A = 3.14 \times 16$$

$$A \approx 50.24$$

The area of the circle is approximately 50.24 square centimeters.

Find the Area of an Ellipse

An ellipse's area is found using its two axes, the major axis (length from the center) usually designated as a, and the minor axis (width from the center), usually designated as b, with this formula:

$A = \pi ab$

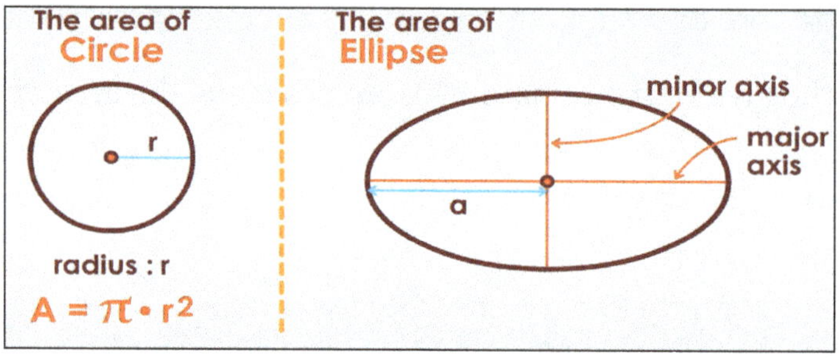

Whether you are dealing with a regular polygon or an irregular plane figure, you can find the area.

Volume

Volume is a measure of how much space an object takes up. For example two shoe boxes together have twice the volume of a single box, because they take up twice the amount of space. Different shapes have different ways to find the volume. For example, in a cube we find the volume by multiplying the three side lengths together. In the cube above, the volume is 3 × 3 × 3 or 27. If you count the small cubes you will find there are 27 of them.

Units of Volume

Given that volume has 3 dimensions, it has a cubic measure of length.

Unit of Side	Unit of Volume
cm	cubic cm or cm^3
mm	cubic mm or mm^3
m	cubic m or m^3

However, while the standard unit of measurement internationally is cubic metre or cubic centimeter, colloquially the most used term is liters or milliliter.

$1\,m^3 = 1{,}1000\,L$

$= 1{,}000{,}000\,cm^3$

$1\,L = 1{,}000\,mL$

$= 1{,}000\,cm^3$

$1\,mL = 1\,cm^3$

Now that we are familiar with the units, let's take a look at how to calculate the volume for a few common shapes.

Volume of Cuboid

Cuboid is a solid box whose every surface is a rectangle of same area or different areas. A cuboid will have a length, breadth and height. Hence we can conclude that volume is 3 dimensional. To measure the volumes we need to know the measure 3 sides. Since volume involves 3 sides it is measured in cubic units.

Volume of a cuboid = (length × breadth × height) cubic units.

= (l × b × h) cubic units.

(Since area = ℓ × b)

Volume of a cuboid = area of one surface × height cubic units.

Let us look at the given cuboid:

- The length of the cuboid = 5 cm.
- The breadth of the cuboid = 3 cm.
- The height of cuboid (thickness) = 2 cm.
- The number of 1 cm cubes in the given cuboid = 30 cubes = 5 × 3 × 2.

We find that volume of the given cuboid with length 5 cm, breadth 3 cm and height 2 cm is 30 cu cm. Therefore, volume of a cuboid = length × breadth × height.

Solved examples on volume of a cuboid:

1) Find the volume of a cuboid of dimensions 14 cm × 12 cm × 8 cm.

Solution: Volume of cuboid = length × breadth × height.

Here, length = 14 cm, breadth = 12 cm and height = 8 cm.

Volume of cuboid = 14 × 12 × 8 cubic cm.

= 1344 cubic cm.

Therefore, volume of cuboid = 1344 cubic cm.

2) Michael made a shoe box with length 8 cm, breadth 6 cm and height 6 cm. Find the volume of the box.

Solution: Volume of the shoe box = Length × breadth × height.

= 8 × 6 × 6.

= 288 cu cm.

3) A fish tank is 40 cm long, 15 cm broad and 10 cm high. What is its volume in cu cm?

Solution:

The length of the fish tank = 40 cm.

The breadth of the fish tank = 15 cm.

The height of the fish tank = 10 cm.

Therefore, the volume of the fish tank = length × breadth × height.

= 40 × 15 × 10 cu. cm.

= 6000 cu cm.

4) Find the volume of a cuboid of dimensions 14 cm × 50 mm × 10 cm.

Solution: Here, length = 14 cm,

[Given, breadth = 50 mm; we need to convert breadth to same unit and then solve. We know, 10 mm = 1 cm. Therefore, 50 mm = 50/10 cm = 5 cm].

Breadth = 5 cm,

Height = 10 cm.

Volume of cuboid = length × breadth × height.

= 14 × 5 × 10.

= 700 cubic cm.

Therefore, volume of cuboid = 700 cubic cm.

In a cuboid, when the length, breadth and height are of different units, convert them to a same unit and then solve.

5) Find the volume of a cuboid of dimensions 17 mm × 0.2 cm × 12 mm in cu. cm.

Solution: Given, length = 17 mm.

We know, 10 mm = 1 cm.

= 17/10 cm.

= 1.7 cm.

Therefore, length = 1.7 cm.

Similarly, height = 12 mm.

We know, 10 mm = 1 cm.

= 12/10 cm.

= 1.2 cm.

Therefore, height = 1.2 cm.

Volume of cuboid = length × breadth × height.

Length = 1.7 cm, breadth = 0.2 cm and height = 1.2 cm.

= 1.7 × 0.2 × 1.2 cu. cm.

= 0.408 cu. cm.

Therefore, volume of cuboid = 0.408 cubic cm.

6) Find the number of cubical boxes of cubical side 3 cm which can be accommodated in carton of dimensions 15 cm × 9 cm × 12 cm.

Solution: Volume of box = side × side × side.

= 3 × 3 × 3.

= 27 cu. cm.

Volume of carton = length × breadth × height.

= 15 × 9 × 12

= 1620 cu. cm.

Number of boxes = Volume of carton/Volume of each box.

= 1620/27.

= 60

Therefore, number of cubical boxes = 60.

7) How many bricks each 25 cm long, 10 cm wide and 7.5 cm thick will be required for a wall 20 m long, 2 m high and 0.75 m thick? If bricks sell at $900 per thousand what will it cost to build the wall?

Solution: Volume of the wall = 20 m × 2 m × 0.75 m.

= 20 × 100 cm × 2 × 100 cm × 0.75 × 100 cm.

Volume of brick = 25 cm × 10 cm × 7.5 cm.

Number of bricks = Volume of the wall/Volume of the brick,

= 20 × 100 × 2 × 100 × 0.75 × 100/25 × 10 × 7.5.

= 16000

The number of bricks = 16000.

The cost of 1 thousand bricks = $ 900.

The cost of building the wall = $ 900 × 16 = $ 14400.

While calculating the volume of a cuboid all the dimensions should be changed into the same unit.

Volume of Cylinder

The volume of a cylinder is the density of the cylinder which signifies the amount of material it can carry or how much amount of any material can be immersed in it. Cylinder's volume is given by the formula, $\pi r^2 h$, where r is the radius of the circular base and h is the height of the cylinder. The material could be a liquid quantity or any substance which can be filled in the cylinder uniformly.

The cylinder is a three-dimensional shape having a circular base. A cylinder can be seen as a set of circular disks that are stacked on one another. Now, think of a scenario where we need to calculate the amount of sugar that can be accommodated in a cylindrical box.

In other words, we mean to calculate the capacity or volume of this box. The capacity of a cylindrical box is basically equal to the volume of the cylinder involved. Thus, the volume of a three-dimensional shape is equal to the amount of space occupied by that shape.

Volume of a Cylinder Formula

A cylinder can be seen as a collection of multiple congruent disks, stacked one above the other. In order to calculate the space occupied by a cylinder, we calculate the space occupied by each disk and then add them up. Thus, the volume of the cylinder can be given by the product of the area of base and height.

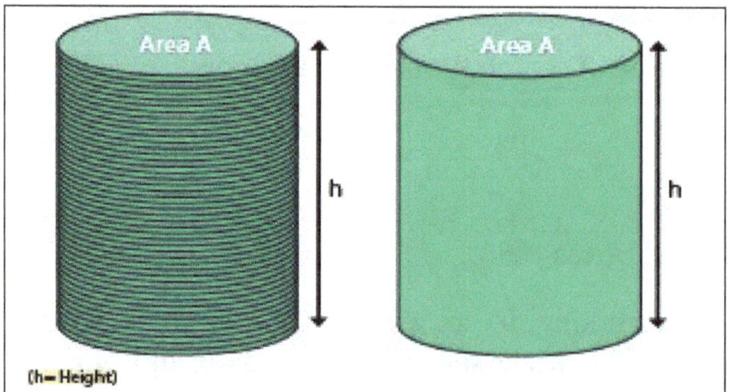

Volume= Base Area x Height = $\pi r^2 h$

For any cylinder with base radius 'r', and height 'h', the volume will be base times the height.

Therefore, the cylinder's volume of base radius 'r', and height 'h' = (area of base) × height of the cylinder.

Since the base is the circle, it can be written as,

Volume = $\pi r^2 \times h$.

Therefore, the volume of a cylinder = $\pi r^2 h$ cubic units.

Volume of Hollow Cylinder

In case of hollow cylinder, we measure two radius, one for inner circle and one for outer circle formed by the base of hollow cylinder. Suppose, r1 and r2 are the two radii of the given hollow cylinder with 'h' as the height, then the volume of this cylinder can be written as:

$$V = \pi h(r_1^2 - r_2^2)$$

Surface Area of Cylinder

The amount of square units required to cover the surface of the cylinder is the surface area of the cylinder. The formula for the surface area of the cylinder is equal to the total surface area of the bases of the cylinder and surface area of its sides.

$$A = 2\pi r^2 + 2\pi rh$$

Volume of Cylinder in Litres

When we find the volume of the cylinder in cubic centimetres, we can convert the value in litres by knowing the below conversion, i.e., 1 Litre = 1000 cubic cm or cm^3. For example: If a cylindrical tube has a volume of 12 litres, then we can write the volume of the tube as $12 \times 1000\ cm^3 = 12{,}000\ cm^3$.

Examples:

Question: Calculate the volume of a given cylinder having height 20 cm and base radius of 14 cm. (Take pi = 22/7).

Solution:

Given:

Height = 20 cm

Radius = 14 cm

We know that;

Volume, $V = \pi r^2 h$ cubic units:

$V = (22/7) \times 14 \times 14 \times 20$

$V = 12320\ cm^3$

Therefore, the volume of a cylinder = $12320\ cm^3$.

Question: Calculate the radius of the base of a cylindrical container of volume $440\ cm^3$. Height of the cylindrical container is 35 cm. (Take pi = 22/7).

Solution:

Given:

Volume = $440\ cm^3$

Height = 35 cm

We know from the formula of cylinder,

Volume, $V = \pi r^2 h$ cubic units,

So, $440 = (22/7) \times r^2 \times 35$

$r^2 = (440 \times 7)/(22 \times 35) = 3080/770 = 4$

Therefore, r = 2 cm.

Therefore, the radius of a cylinder = 2 cm.

Volume of Cone

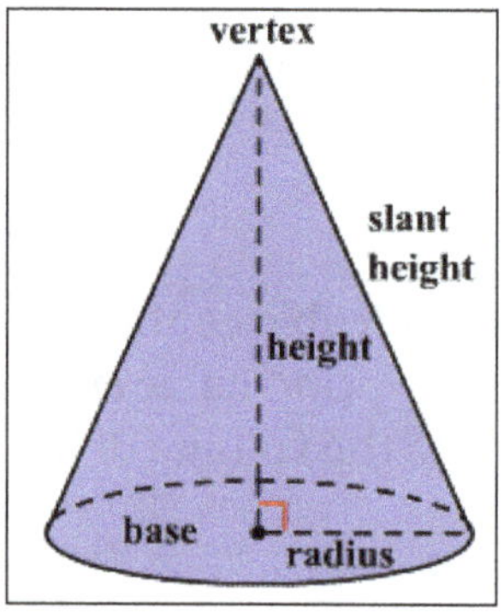

A cone is a three-dimensional figure with one circular base. A curved surface connects the base and the vertex. The volume of a 3-dimensional solid is the amount of space it occupies. Volume is measured in cubic units (in^3, ft^3, cm^3, m^3 et cetera). Be sure that all of the measurements are in the same unit before computing the volume.

The volume V of a cone with radius r is one-third the area of the base B times the height h.

$$V = \frac{1}{3}Bh \text{ or } V = \frac{1}{3}\pi r^2 h, \text{ where } B = \pi r^2$$

The formula for the volume of an oblique cone is the same as that of a right one.

The volumes of a cone and a cylinder are related in the same way as the volumes of a pyramid and a prism are related. If the heights of a cone and a cylinder are equal, then the volume of the cylinder is three times as much as the volume of a cone.

Example: Find the volume of the cone shown. Round to the nearest tenth of a cubic centimeter.

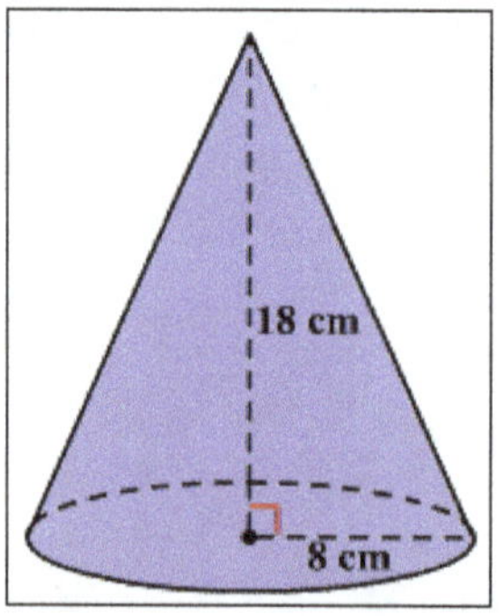

Solution: From the figure, the radius of the cone is 8 cm and the height is 18 cm.

The formula for the volume of a cone is,

$$V = \frac{1}{3}\pi r^2 h$$

Substitute 8 for r and 18 for h.

$$V = \frac{1}{3}\pi(8)^2(18)$$

Simplify:

$$V = \frac{1}{3}\pi(64)(18)$$
$$= 384\pi$$
$$\approx 1206.4$$

Therefore, the volume of the cone is about 1206.4 cubic centimeters.

Volume of Prism

The volume of a prism is total space occupied by a prism. A prism is a geometric solid figure with two identical ends, flat faces and the same cross section all along its length. Prisms are named after the shapes of their cross section. For, example, a prism with a triangular cross section is known as triangular prism. Other examples of prisms include rectangular prism, pentagonal prism, hexagonal prism, trapezoidal prism etc.

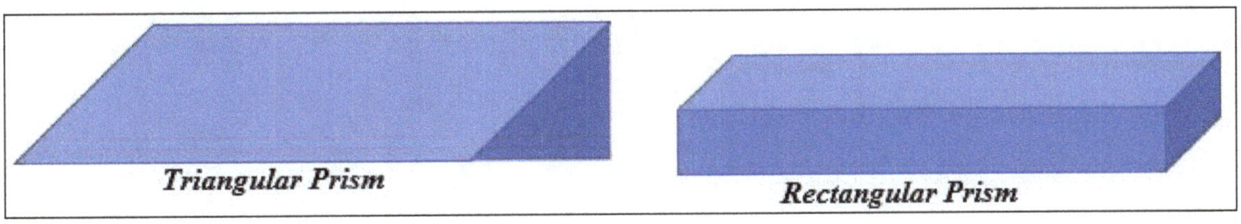

To find the volume of a prism, you require the area and the height of a prism. The volume of a prism is calculated by multiplying the base area and the height. The volume of a prism is also measured in cubic units i.e. cubic meters, cubic centimeters etc.

Volume of a Prism Formula

The formula for calculating the volume of a prism depends on the cross section or base of a prism. Since we already know the formula for calculating the area of polygons, finding the volume of a prism is as easy as a pie. The general formula for the volume of a prism is given as,

The volume of a Prism = Base Area × Length

Where, Base is the shape of a polygon that is extruded to form a prism.

Volume of a Triangular Prism

A triangular prism is a prism whose cross section is a triangle. The formula for the volume of a triangular prism is given as,

Volume of a triangular prism = ½ abh

where,

a = apothem of a triangular prism.

The apothem of a polygon is the line connecting the center of the polygon to the midpoint of one of the polygon's side. The apothem of a triangle is the height of a triangle.

b = base length of a triangle,

h = height of a prism.

Example: Find the volume of a triangular prism whose apothem is 12 cm, base length is 16 cm and height, is 25 cm.

Solution: By the formula of a triangular prism,

volume = ½ abh

= ½ x 12 x 16 x 25

= 150 cm^3

Example: Find the volume of a prism whose height is 10 cm and the cross section is an equilateral triangle of side length 12 cm.

Solution: Find the apothem of the triangular prism.

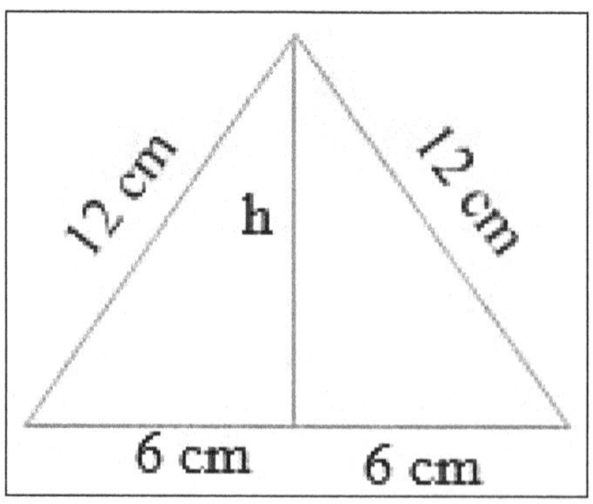

By Pythagorean theorem,

$h^2 + 6^2 = 12^2$

$h^2 + 36 = 144$

$h^2 = 108$

h = 10.4 cm

Therefore, the apothem of the prism is 10.4 cm.

Volume = ½ abh

= ½ x 10.4 x 12 x 10

= 624 cm^3

Volume of a Pentagonal Prism

For a pentagonal prism, the volume is given by formula:

Volume of a pentagonal prism = (5/2) abh

Where,

a = apothem of a pentagon,

b = base length of a pentagonal prism,

h = height of a prism.

Example: Find the volume of a pentagonal prism whose apothem is 10 cm, base length is 20 cm and height, is 16 cm.

Solution: Volume of a pentagonal prism = (5/2) abh,

= (5/2) x 10 x 20 x 16

= 8000 cm^3

Volume of a Hexagonal Prism

A hexagonal prism has a hexagon as the base or cross section. The volume of a hexagonal prism is given by:

Volume of a hexagonal prism = 3abh

where,

a = apothem length of a hexagon,

b = base length of a hexagonal prism,

h = height of a prism.

Example: Calculate the volume of a hexagonal prism with the apothem as 5 m, base length as 12 m and height as 6 m.

Solution: Volume of a hexagonal prism = 3abh

= 3 x 5 x 12 x 6

= 1080 m^3

Alternatively, if the apothem of a prism is not known, then the volume of any prism is calculated as follows:

Volume of a prism = (h)(n) (s^2)/ [4 tan (180/n)]

Where, h = height of a prism

s = side length of the extruded regular polygon,

n = number of sides of a polygon,

tan = tangent.

This formula is only applied where the base or the cross section of a prism is a regular polygon.

Example: Find the volume of a pentagonal prism with a height of 0.3 m and side length of 0.1 m.

Solution: In this case, n = 5,

h = 0.3 m and s = 0.1 m

By substitution,

Volume of a pentagonal prism = (0.3) (5) (0.1^2)/ [4 tan (180/5)],

= 0.015/4 tan 36,

= 0.015/2.906,

= 0.00516 m^3.

References

- Axioms-conjectures-and-theorems, introduction-to-euclids-geometry, maths: toppr.com, Retrieved 11, June 2020
- An-introduction-to-geometry, points,-lines,-planes-and-angles, geometry, education: mathplanet.com, Retrieved 25, May 2020
- Types-of-angles: storyofmathematics.com, Retrieved 16, April 2020
- Perpendicular-from-point-to-line, geometry: cuemath.com, Retrieved 26, March 2020
- What-is-area-in-math-definition-formula, geometry-help, math-tutors: tutors.com, Retrieved 27, July 2020

3

Polygons and their Classification

A polygon may be defined as a flat, two-dimensional closed shape with straight sides. Polygons come in all shapes and number of sides. Regular polygons, irregular polygons, convex polygons, concave polygons, quadrilateral polygons, pentagon polygons, etc. are a few examples. This chapter has been carefully written to provide the reader with a clearer understanding of polygons and their different types.

A polygon is a plane figure that closes in a space using only line segments. If it must use only line segments and must close in a space, the polygon with the fewest sides has to be the triangle (three sides and interior angles). Polygons have no upper limit to the number of sides and interior angles, so mathematicians say n-gon after the familiar names we use for common polygons.

Table: Names of Polygons.

Polygon Shape	Number of Sides
Triangle	3 sides
Square	4 sides
Rectangle	4 sides
Quadrilateral	4 sides
Parallelogram	4 sides
Rhombus	4 sides
Dart	4 sides
Kite	4 sides
Pentagon	5 sides
Hexagon	6 sides
Heptagon	7 sides
Octagon	8 sides
Nonagon	9 sides
Decagon	10 sides
Dodecagon	12 sides
Icosagon	20 sides
Hectagon	100 sides
n-gon	n sides

Polygon Shapes

Let's take a look at the vast array of shapes that are polygons.

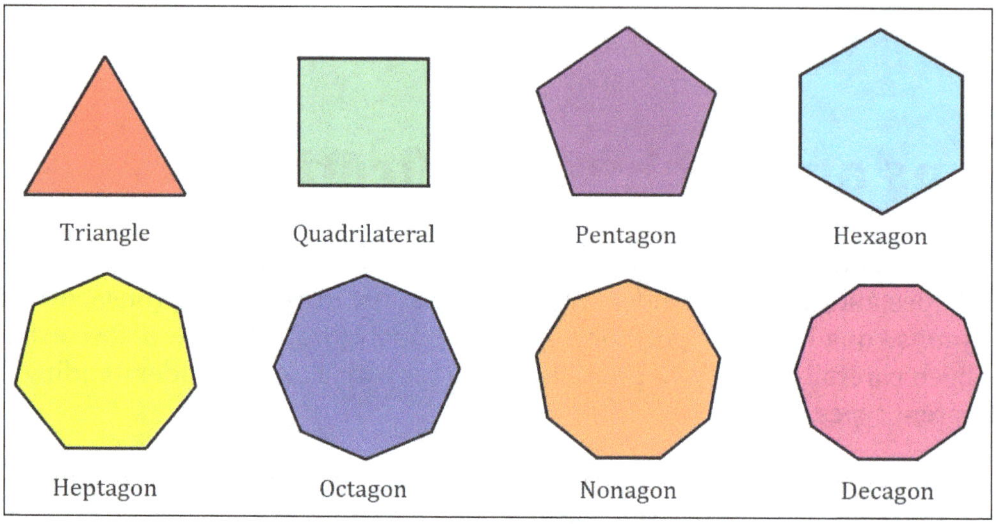

Regular Polygon

A regular polygon is an n-sided polygon in which the sides are all the same length and are symmetrically placed about a common center (i.e., the polygon is both equiangular and equilateral). Only certain regular polygons are "constructible" using the classical Greek tools of the compass and straightedge.

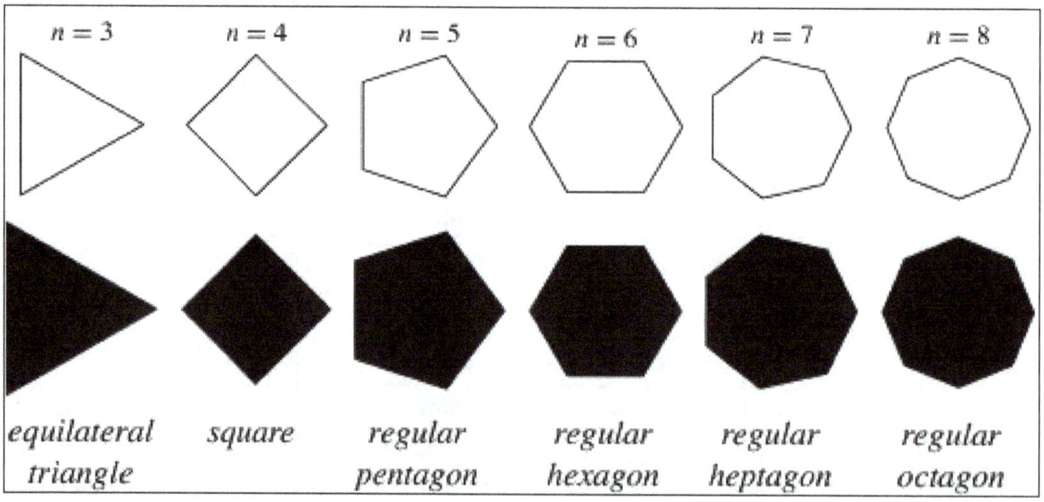

The terms equilateral triangle and square refer to the regular 3- and 4-polygons, respectively. The words for polygons with $n \geq 5$ sides (e.g., pentagon, hexagon, heptagon, etc.) can refer to either regular or non-regular polygons, although the terms generally refer to regular polygons in the absence of specific wording.

A regular n-gon is implemented in the Wolfram Language as RegularPolygon[n], or more generally as RegularPolygon[r, n], RegularPolygon[{x, y}, rspec, n], etc. The sum of perpendiculars from any point to the sides of a regular polygon of n sides is n times the apothem.

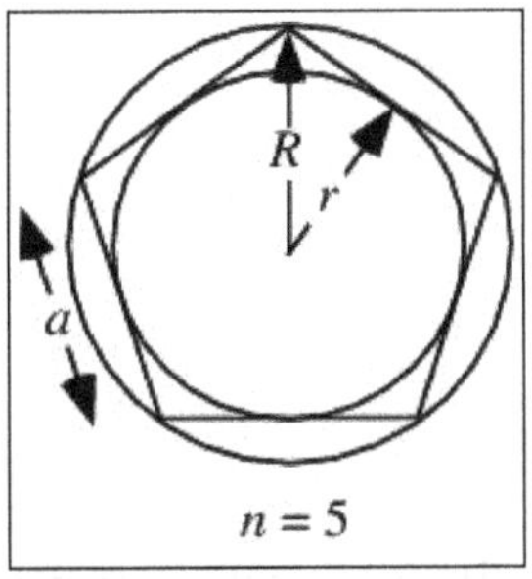

Let a be the side length, r be the in radius, and R the circumradius of a regular polygon. Then,

$$a = 2r\tan\left(\frac{\pi}{n}\right)$$
$$= 2R\sin\left(\frac{\pi}{n}\right)$$
$$r = \frac{1}{2}a\cot\left(\frac{\pi}{n}\right)$$
$$= R\cos\left(\frac{\pi}{n}\right)$$
$$R = \frac{1}{2}a\csc\left(\frac{\pi}{n}\right)$$
$$= r\sec\left(\frac{\pi}{n}\right)$$
$$A = \frac{1}{4}na^2\cot\left(\frac{\pi}{n}\right)$$
$$= nr^2\tan\left(\frac{\pi}{n}\right)$$
$$= \frac{1}{2}nR^2\sin\left(\frac{2\pi}{n}\right).$$

The area moments of inertia about axes along an inradius and a circumradius of a regular n-gon are given by,

$$I_r = \frac{1}{24}A_n\left(6r_n^2 - a^2\right)$$
$$= \frac{a^4}{192}n\left[\cos\left(\frac{2\pi}{n}\right) + 2\right]\cos\left(\frac{\pi}{n}\right)\csc^2\left(\frac{\pi}{n}\right)$$

$$I_R = \frac{1}{48}A_n\left(12R_n^2 + a^2\right)$$
$$= \frac{a^4}{192}n\cot\left(\frac{\pi}{n}\right)\left[3\cos^2\left(\frac{\pi}{n}\right) + 1\right]$$

If the number of sides is doubled, then,

$$a_{2n} = \sqrt{2R^2 - R\sqrt{4R^2 - a_n^2}}$$

$$A_{2n} = \frac{4rA_n}{2r + \sqrt{4r^2 + a_n^2}}.$$

The area of the first few regular n-gon with unit edge lengths are,

$$A_3 = \frac{1}{4}\sqrt{3}$$

$$A_4 = 1$$

$$A_5 = \frac{1}{4}\sqrt{5\left(5+2\sqrt{5}\right)}$$

$$A_6 = \frac{3}{2}\sqrt{3}$$

$$A_7 = \left(4096x^6 - 62720x^4 + 115248x^2 - 16807\right)_6$$

$$A_8 = 2\left(1+\sqrt{2}\right)$$

$$A_9 = \left(4096x^6 - 62720x^4 + 115248x^2 - 16807\right)_6$$

$$A_{10} = \frac{5}{2}\sqrt{5+2\sqrt{5}}.$$

The algebraic degrees of these for n = 3, 4, ... are 2, 1, 4, 2, 6, 2, 6, 4, 10, 2, 12, 6, 8, 4, 16, 6, 18, 4.

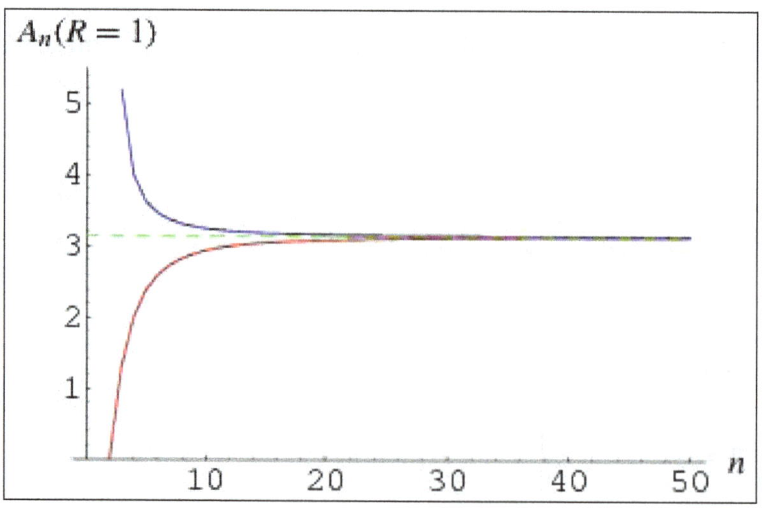

The plot above shows how the areas of the regular n-gons with unit inradius (blue) and unit circumradius (red) approach that of a unit disk (i.e., π).

If p_k and P_k are the perimeters of the regular polygons inscribed in and circumscribed around a given circle and a_k and A_k their areas, then,

$$P_{2n} = \frac{2p_n P_n}{p_n + P_n}$$

$$p_{2n} = \sqrt{p_n P_{2n}},$$

and

$$a_{2n} = \sqrt{a_n A_n}$$

$$A_{2n} = \frac{2a_{2n} A_n}{a_{2n} + A_n}$$

The sum of interior angles in any n-gon is given by $(n - 2)\pi$ radians, or $2(n - 2) \times 90°$.

The following table gives parameters for the first few regular polygons of unit edge length $s = 1$, where α is the interior (vertex) angle, β is the exterior angle, r is the inradius, R is the circumradius, and A is the area,

Polygon	$\{n\}$	α	β	r	R	A
Equilateral triangle	$\{3\}$	$\frac{1}{3}\pi = 60°$	$\frac{2}{3}\pi = 120°$	$\frac{1}{6}\sqrt{3}$	$\frac{1}{3}\sqrt{3}$	$\frac{1}{4}\sqrt{3}$
Square	$\{4\}$	$\frac{1}{2}\pi = 90°$	$\frac{1}{2}\pi = 90°$	$\frac{1}{2}$	$\frac{1}{2}\sqrt{2}$	1
Pentagon	$\{5\}$	$\frac{3}{5}\pi = 108°$	$\frac{2}{5}\pi = 72°$	$\frac{1}{10}\sqrt{25+10\sqrt{5}}$	$\frac{1}{10}\sqrt{50+10\sqrt{5}}$	$\frac{1}{4}\sqrt{25+10\sqrt{5}}$
Hexagon	$\{6\}$	$\frac{2}{3}\pi = 120°$	$\frac{1}{3}\pi = 60°$	$\frac{1}{2}\sqrt{3}$	1	$\frac{3}{2}\sqrt{3}$
Heptagon	$\{7\}$	$\frac{5}{7}\pi = \frac{900}{7}°$	$\frac{2}{7}\pi = \frac{360}{7}°$	$\frac{1}{2}\cot\left(\frac{1}{7}\pi\right)$	$\frac{1}{2}\csc\left(\frac{1}{7}\pi\right)$	$\frac{7}{4}\cot\left(\frac{1}{7}\pi\right)$
Octagon	$\{8\}$	$\frac{3}{4}\pi = 135°$	$\frac{1}{4}\pi = 45°$	$\frac{1}{2}\left(1+\sqrt{2}\right)$	$\frac{1}{2}\sqrt{4+2\sqrt{2}}$	$2\left(1+\sqrt{2}\right)$
Nonagon	$\{9\}$	$\frac{7}{9}\pi = 140°$	$\frac{2}{9}\pi = 40°$	$\frac{1}{2}\cot\left(\frac{1}{9}\pi\right)$	$\frac{1}{2}\csc\left(\frac{1}{9}\pi\right)$	$\frac{9}{4}\cot\left(\frac{1}{9}\pi\right)$

Decagon	$\{10\}$	$\frac{4}{5}\pi = 144°$	$\frac{1}{5}\pi = 36°$	$\frac{1}{2}\sqrt{5+2\sqrt{5}}$	$\frac{1}{2}\left(1+\sqrt{5}\right)$	$\frac{5}{2}\sqrt{5+2\sqrt{5}}$
Hendecagon	$\{11\}$	$\frac{9}{11}\pi = \frac{1620}{11}°$	$\frac{2}{11}\pi = \frac{360}{11}°$	$\frac{1}{2}\cot\left(\frac{1}{11}\pi\right)$	$\frac{1}{2}\csc\left(\frac{1}{11}\pi\right)$	$\frac{11}{4}\cot\left(\frac{1}{11}\pi\right)$
Dodecagon	$\{12\}$	$\frac{5}{6}\pi = 150°$	$\frac{1}{6} = 30°$	$\frac{1}{2}\left(2+\sqrt{3}\right)$	$\frac{1}{2}\left(\sqrt{2}+\sqrt{6}\right)$	$3\left(2+\sqrt{3}\right)$
Tridecagon	$\{13\}$	$\frac{11}{13}\pi = \frac{1980}{13}°$	$\frac{2}{13}\pi = \frac{360}{13}°$	$\frac{1}{2}\cot\left(\frac{1}{13}\pi\right)$	$\frac{1}{2}\csc\left(\frac{1}{13}\pi\right)$	$\frac{13}{4}\cot\left(\frac{1}{13}\pi\right)$
Tetradecagon	$\{14\}$	$\frac{6}{7}\pi = \frac{1080}{11}°$	$\frac{1}{7}\pi = \frac{180}{7}°$	$\frac{1}{2}\cot\left(\frac{1}{14}\pi\right)$	$\frac{1}{2}\csc\left(\frac{1}{14}\pi\right)$	$\frac{7}{2}\cot\left(\frac{1}{14}\pi\right)$

Only some of the regular polygons can be built by geometric construction using a compass and straightedge. The numbers of sides for which regular polygons are constructible are those having central angles corresponding to so-called trigonometry angles.

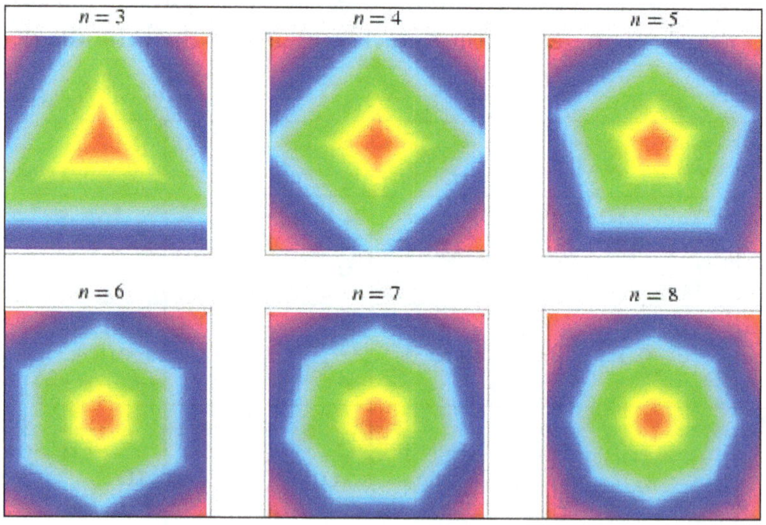

It is possible to construct relatively simple two-dimensional functions $P_n(x, y)$ that have the symmetry of a regular n-gon (i.e., whose level curves are regular n-gons). Examples, illustrated above, include,

$$P_3(x,y) = \max\left(y - x\sqrt{3},\ y + x\sqrt{3}, -2y\right)$$

$$P_4(x,y) = |x| + |y|$$

$$P_6(x,y) = 2|x| + \left|x - y\sqrt{3}\right| + \left|x + y\sqrt{3}\right|$$

$$P_8(x,y) = 2\left(|x| + |y|\right) + \sqrt{2}\left(|x - y| + |x + y|\right).$$

Irregular Polygon

An irregular polygon is any polygon that is not a regular polygon. It can have sides of any length and each interior angle can be any measure. They can be convex or concave, but all concave polygons are irregular since the interior angles cannot all be the same. If you drew a polygon at random, it would probably be irregular.

Equilateral Polygons

Equilateral means that all sides are the same length. But as can be seen from the figure below, this can still be irregular. This figure has all sides the same length, but it does not have all angles the same, so it is irregular.

Finding the Area of an Irregular Polygon

As with any kind of polygon, finding the perimeter is a relatively trivial exercise. Providing you know the length of each side, you can simply add the lengths together to find the perimeter. If the lengths of some or all of the sides are unknown, you will first need to find the length of each side by measuring it. Finding the area is more complicated. In the case of irregular polygons that are either triangles or quadrilaterals. For irregular polygons that have more than four sides, other methods must be used. The exact method used in any given case will depend on the kind of irregular polygon we are dealing with. Consider the irregular hexagon illustrated below, for example:

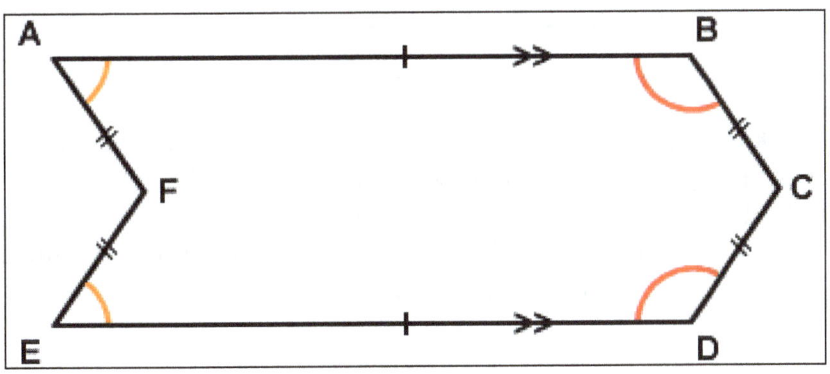

An irregular hexagon.

In this case, we know that the line segments AB and DE are equal in length, as are the line segments BC, CD, EF and FA. We also know that the line segments AB and DE are parallel. From this information, we can deduce that angles ABC and CDE are equal, and that angles DEF and FAB are equal. We can also deduce that angles ABC and FAB are supplementary (add up to one hundred and eighty degrees) as are angles CDE and DEF. It can also be shown that the line segments BC and FA are parallel, as are the line segments CD and EF. From this information, you should be able to see that if we were to take the triangular portion of the shape defined by points B, C and D and move it so that it covered the triangular area defined by points A, E and F, we would be left with a rectangle as shown.

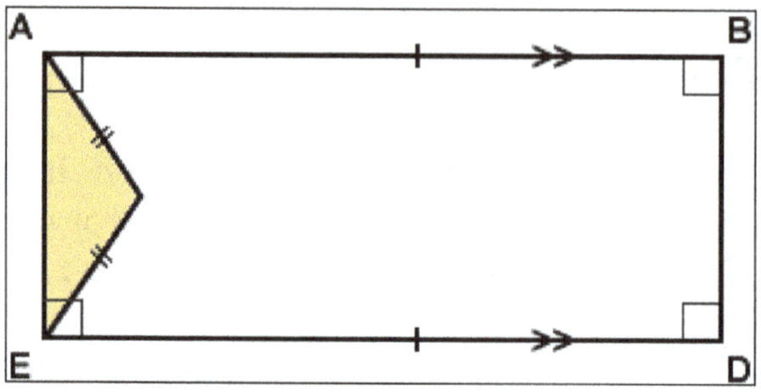

The irregular hexagon becomes a rectangle.

We can see that the irregular hexagon ABCDEF is equivalent in area to the rectangle ABCD, so the area of ABCDEF can be found by taking the product of the lengths of line segments AB and BD. If these lengths are known, this is a straightforward calculation. If not, they must first be found by measurement. In cases such as the above, it is often possible to transform the irregular polygon into a shape for which we already have a method of calculating the area, and which has the same area as the original shape. In other cases this is either difficult or not possible. We could, alternatively, have drawn the line segment FC as shown below, effectively breaking the shape into two parallelograms of equal area. We could then find the area of one of the parallelograms and multiply the result by two to find the total area of the shape.

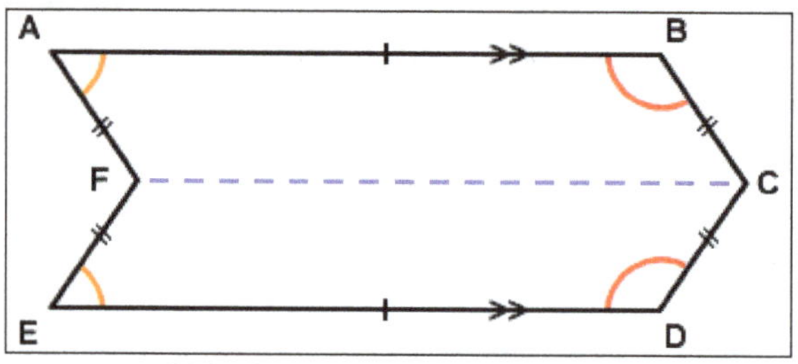

The irregular hexagon could be broken down into two parallelograms of equal area.

One method that works for finding the area of any irregular polygon (or any regular polygon for that matter) involves breaking the polygon down into triangles, finding the area of each triangle using standard methods, and adding the areas of the individual triangles together (note that in some cases, a shape can be broken down into a combination of triangles and rectangles, but all polygons can be broken down into triangles). The principle is illustrated below:

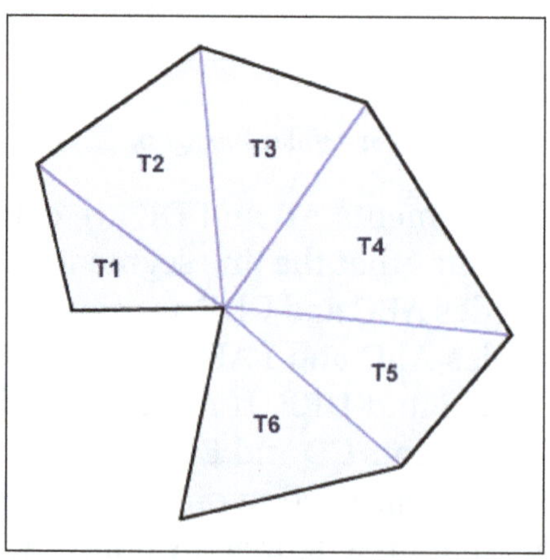

Any polygon may be broken down into a number of triangular areas.

The area of the polygon is the sum of the areas of triangles T_1, T_2, T_3, T_4, T_5 and T_6. The main disadvantage of this method is that it will involve selecting one side of each triangle to be the base, constructing the perpendicular line segment from the base to the apex of the triangle (i.e. the angle opposite the base), and measuring the dimensions of each. Once you know the length l of the base and the height h for each triangle it is a straightforward calculation to find the area:

$$\text{Area} = \frac{l \times h}{2}$$

There is also a formula that can be used to find the area of a simple convex or concave irregular polygon, even if every side is of a different length and every angle of a different magnitude. The formula can only be used, however, if you know the coordinates of each of the vertices. It also has the disadvantage that it cannot be used for complex polygons (i.e. polygons in which two or more of the sides intersect one another). Consider the following concave irregular polygon:

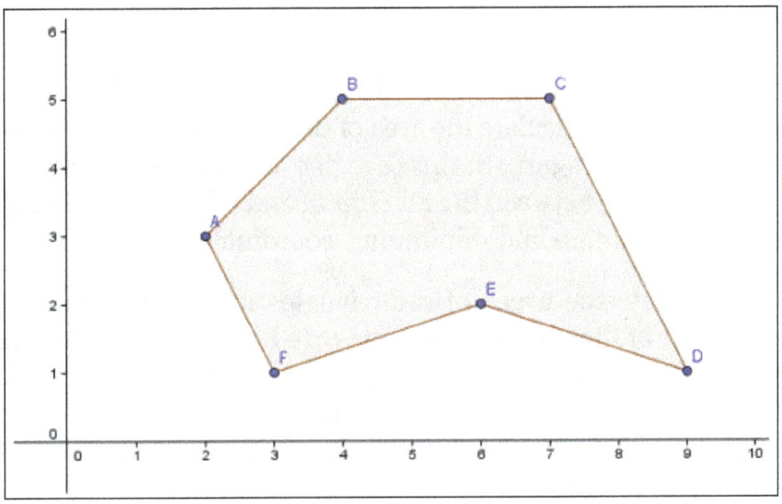

An irregular concave six-sided polygon.

The coordinates of the vertices for the irregular six-sided figure shown above are as follows:

A 2, 3

B 4, 5

C 7, 5

D 9, 1

E 6, 2

F 3, 1

The general formula used to find the area of a simple polygon uses the xy coordinates of each vertex of the polygon from the first to the last in clockwise order around the shape as follows:

$$\text{Area} = \frac{(x_1y_2 - y_1x_2) + (x_2y_3 - y_2x_3) \ldots + (x_ny_1 - y_nx_1)}{2}$$

Applying the formula to our six-sided irregular polygon, we get:

$$\text{Area} = \frac{(x_A y_B - y_A x_B) + (x_B y_C - y_B x_C) + (x_C y_D - y_C x_D) + (x_D y_E - y_D x_E) + (x_E y_F - y_E x_F) + (x_F y_A - y_F x_A)}{2}$$

$$\text{Area} = \frac{(2\cdot 5 - 3\cdot 4) + (4\cdot 5 - 5\cdot 7) + (7\cdot 1 - 5\cdot 9) + (9\cdot 2 - 1\cdot 6) + (6\cdot 1 - 2\cdot 3) + (3\cdot 3 - 1\cdot 2)}{2}$$

$$\text{Area} = \frac{(10 - 12) + (20 - 35) + (7 - 45) + (18 - 6) + (6 - 6) + (9 - 2)}{2}$$

$$\text{Area} = \frac{-2 - 15 - 38 + 12 + 0 + 7}{2}$$

$$\text{Area} = \frac{-36}{2} = -18$$

Don't worry if the result turns out to be negative. Simply change the minus sign to a plus sign. As a quick check on the result, you can calculate the area of the bounding rectangle, since the answer you get should always be less than this figure. In this case, the area of the bounding rectangle will be the product of 9 - 2 (i.e. the difference between the maximum and minimum x coordinates) and 5 - 1 (i.e. the difference between the maximum and minimum y coordinates), which gives us 7 × 4 = 28.

It may be possible to calculate the areas of the triangles formed between the perimeter of the bounding rectangle and that of the polygon, as is the case here. Three of the triangles so formed in the example we have used are right-angled triangles. For these triangles, we know the lengths of the sides adjacent to the right-angle, which essentially gives us a base length and height. We also know both the base length and the height of the remaining triangle (triangle DEF), even though this triangle does not contain a right-angle.

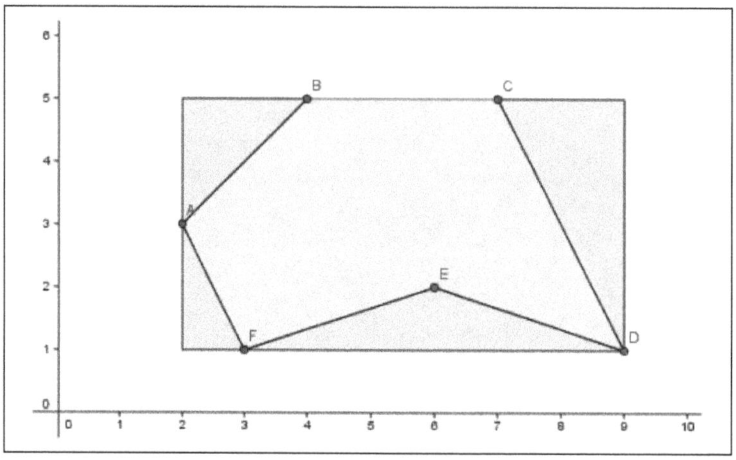

An irregular concave six-sided polygon with its bounding rectangle.

The resulting calculation would look something like this:

$$\text{Area} = (7 \times 4) - \frac{(2 \times 2) + (2 \times 4) + (6 \times 1) + (1 \times 2)}{2}$$

$$\text{Area} = 28 - \frac{20}{2}$$

$$\text{Area} = 28 - 10 = 18$$

Convex Polygon

Convex polygon is a polygon that has all interior angles less than 180°. Convex polygons are polygons for which a line segment joining any two points in the interior lies completely within the figure. This means that all the vertices of the polygon will point outwards, away from the interior of the shape. Think of it as a 'bulging' polygon.

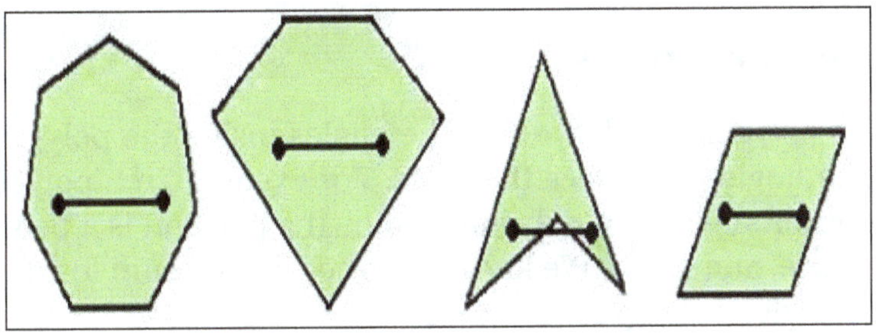

A line drawn through a convex polygon will intersect the polygon exactly twice. You can also see that the line will divide the polygon into exactly two pieces. All the diagonals of a convex polygon lie entirely inside the polygon. The area of an irregular convex polygon can be found by dividing it into triangles and summing the triangle's areas.

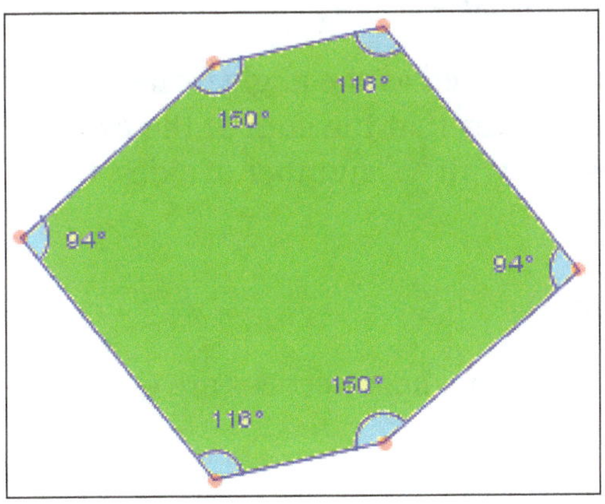

A convex polygon must satisfy the following properties:

- It has all the interior angles less than or equal to 180°.
- It has both the sides joining a vertex either going inside or straight.
- It has each diagonal lying inside or on the boundary of the polygon.
- It has all the vertices pointing outwards.

- If a line segment is drawn through the polygon, then it intersects the boundary of polygon exactly twice.

Convex Polygon Formula

The formula to find the area of a regular convex polygon is given as follows:

If the convex polygon has vertices (x_1, y_1), (x_2, y_2), (x_3, y_3),........, (x_n, y_n), then the formula for the area of the convex polygon is,

$$A = \frac{1}{2}\left|\begin{pmatrix} x_1 & x_2 \\ y_1 & y_2 \end{pmatrix} + \begin{pmatrix} x_2 & x_3 \\ y_2 & y_3 \end{pmatrix} + \text{........} + \begin{pmatrix} x_n & x_1 \\ y_2 & y_1 \end{pmatrix}\right|$$

Convex Polygon Interior Angles

The interior angles of a polygon are defined as the angles inside the polygon. A polygon has the same number of interior angles as it does the sides. For example, if a polygon is a quadrilateral, then the number of interior angles of a polygon is four. If a polygon is a pentagon, then the number of interior angles is five, and so on. We know that the polygon sum formula states that, for any n-polygon, the interior angles sum up to $(n - 2)180°$.

Once if the sum of the interior angles of a convex polygon is known, we can easily find one of its interior angles when all the sides are congruent. It can be found by dividing the sum of the interior angles by the total number of sides.

Convex Polygon Exterior Angles

The exterior angle of a polygon is defined as the angle formed bt extending the sides of a polygon. The exterior angle sum theorem states that the sum of the exterior angles of a convex polygon is 360°. If a convex polygon is regular with "n" number of sides, then each exterior angle of a convex polygon is measured as $360°/n$.

Convex Polygon Example

Question: Find the area of the polygon whose vertices are (5, 7), (9, 2) and (-4, 8).

Solution:

Given: The vertices are: (5, 7), (9, 2) and (-4, 8),

Here, $(x_1, y_1) = (5, 7)$

$(x_2, y_2) = (9, 2)$

$(x_3, y_3) = (-4, 8)$

The formula to find the area of a convex polygon is,

$$A = ½ \mid (x_1y_2 - x_2y_1) + (x_2y_3 - x_3y_2) + (x_3y_1 - x_1y_3) \mid$$

A = ½ | (10-63) + (72 +8) (-40-28)|

A = ½ | -41|

A = ½ |41|

A = 41/2

Therefore, the area of the convex polygon is 41/2.

Concave Polygon

A polygon has at least one angle that measures more than 180 degrees, which is called concave polygon. The vertices (endpoints) of this polygon are inwards as well as outwards. They are just opposite of the convex polygons. A triangle cannot be considered as a concave polygon because it has only three sides and whose sum of interior angles is 180 degrees.

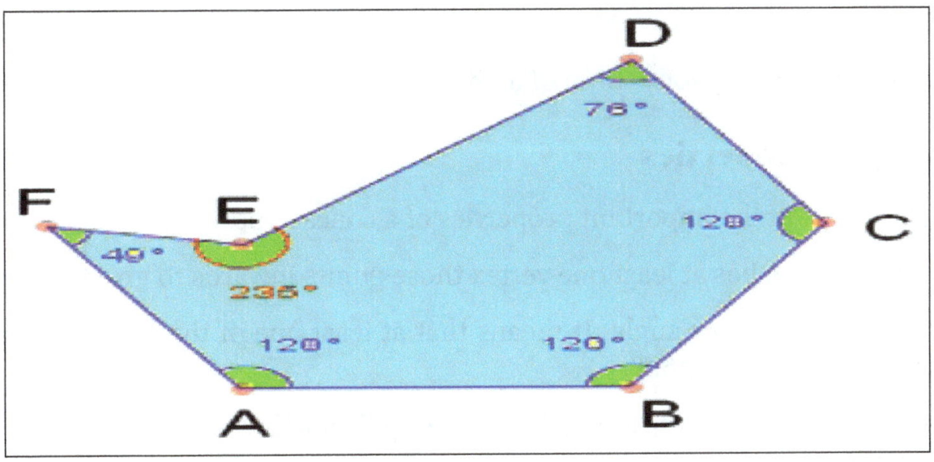

Irregular Concave Polygon

A polygon may be an either convex or concave polygon. Those polygons are further classified into regular or irregular. The irregular polygon can have sides with different measures and also each interior angles measure is also varied. It is noted that all the concave polygons are irregular since the interior angles of the polygon are of different measures. Is Star a Concave Polygon? Yes, a star is a concave polygon because concave polygon should have at least 4 sides. Also, one or more interior angles should be greater than 180 degrees.

Regular Concave Polygon

A regular polygon is a polygon where the length of each side is the same and all the interior angles are equal. By the definition of a concave polygon, it contains at least one of the interior angles more than 180 degrees. Also, the sum of the interior angles of a polygon is (n – 2) x 180, where n is the number of sides. So, it is not possible to have a polygon with all sides equal and an angle greater than 180 degrees. Hence, regular polygons are never concave.

Concave Polygon Formulas

Let us discuss the formulas such as area and the perimeter of the concave polygon below:

Area of a Concave Polygon

Unlike a regular polygon, there is no easy formula to find the area of a concave polygon. Each side could be of a different length, and each interior angle could be different. So, we have to split the concave polygon as triangles or parallelograms or other shapes for which we can easily find the area.

Area of Concave Polygon = Area of the different shapes available in it.

The Perimeter of a Concave Polygon

The perimeter of any polygon is defined as the total distance covered around the boundary of the polygon. Similarly, the perimeter of a concave polygon is defined as the total distance covered around the boundary of the concave polygon. The perimeter of a concave polygon can be found by adding together the length of all the sides.

The perimeter of Concave Polygon = Sum of all its sides.

Concave Polygon Properties

The following are some of the important properties of a concave polygon:

- A concave polygon has at least one vertex those points inwards to give the concave shape.
- It has at least one reflex angle. It means that at least one of the interior angles is greater than 180° and less than 360°.
- If a line segment is drawn crossing the concave polygon, it will intersect the boundary more than two times.
- A polygon can have more than one diagonal that lie outside the boundary.
- A concave polygon has at least one pair of sides joining a vertex that goes outside the vertex.

Exterior Angle of a Concave Polygon

The exterior angles of a polygon always add up to 3600. A concave polygon has one or more of its vertices "pushed inside". Hence, they point towards the interior of the polygon.

Interior Angle of a Concave Polygon

The interior angles of any polygon always add up to a constant value, which depends only on the number of sides of the polygon. For example, the interior angles of a pentagon always add up to 540°, no matter if it is convex or concave, or what size and shape it is. The sum of the interior angles formula of a polygon is given by:

Sum of interior angles = 180 * (n – 2) degrees

where n is the number of sides,

- Square: n = 4; sum of interior angles = 180 x (4 - 2) = 360 degrees.
- Pentagon: n = 5; sum of interior angles = 180 x (5 - 2) = 540 degrees.
- Hexagon: n = 6; sum of interior angles = 180 x (6 - 2) = 720 degrees.

Concave Polygon Example

Question: Find the area and perimeter for the concave polygon given below:

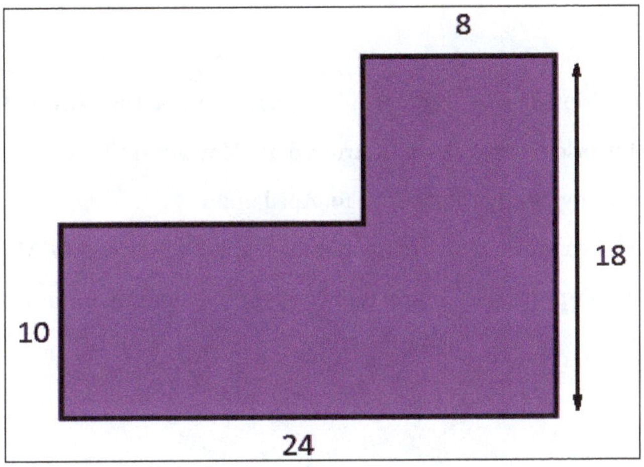

Solution: In this figure, one of the shapes is rectangle and the other one is a square.

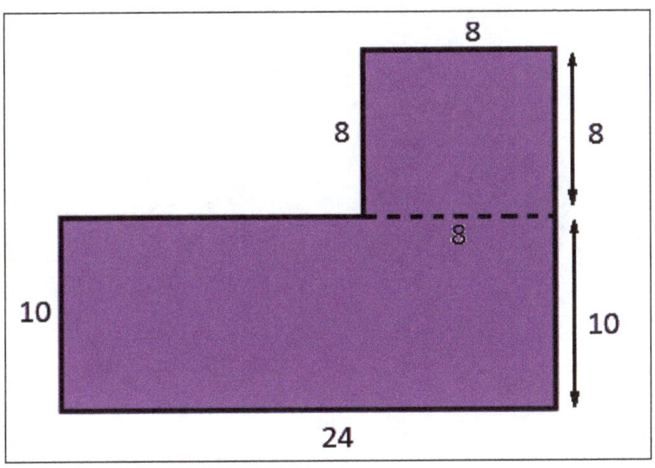

First, find the area of rectangle and square and then add the two areas to get the total area of a concave polygon.

Step 1: Find the area of the rectangle? Area of the rectangle = length x width. Here, length = 24 and width = 10. Area = 24 x 10 = 240 sq units.

Step 2: Find the area of the square? Area of square = Side x Side,

Side = 8Area = 8 x 8 = 64 sq units.

Step 3: Total area of the concave polygon = Area of rectangle + Area of square,

TA = 240 + 64,

TA = 304 sq units.

Step 4: Perimeter of given polygon = Sum of all sides,

P = 24 + 18 + 8 + 8 + 16 + 10,

P = 84 units.

References

- What-is-a-polygon-definition-shapes, geometry-help, math-tutors: tutors.com, Retrieved 10, July 2020
- Regular-Polygon: mathworld.wolfram.com, Retrieved 27, May 2020
- Concave-polygon, maths: byjus.com, Retrieved 19, April 2020
- Irregular-polygons, geometry, mathematics: technologyuk.net, Retrieved 16, May 2020
- Convex-polygon-definition-properties, mathematic, science: assignmentpoint.com, Retrieved 12, June 2020

4

Triangles and its Types

A triangle is a two-dimensional, three-sided closed polygon. Triangles can be sorted into two groups based on their sides and angles. Equilateral triangles, isosceles triangles and scalene triangles are based on their sides. Triangles based on their angles are right triangles, obtuse triangles and acute triangles. This chapter provides a complete overview of triangles for the reader's benefit.

A triangle is composed of three line segments. The line segments intersect in their endpoints. To name a triangle we often use its vertices (the name of the endpoints). The triangle below is named ABC.

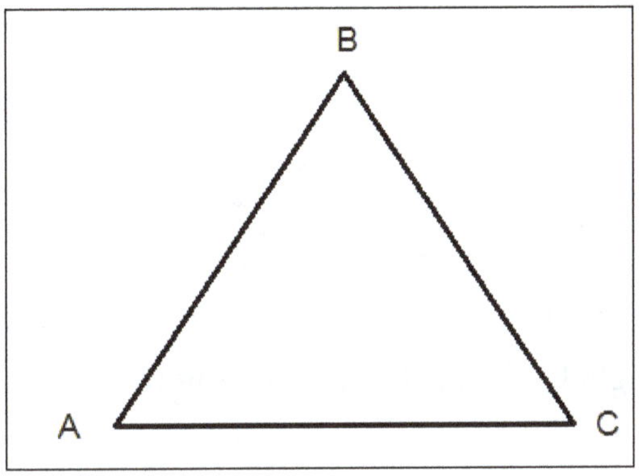

A triangle has three angles. The sum of the measures of the angles is always 180° in a triangle. We have different types of triangles. A triangle is classified by its angles and by the number of congruent sides.

Basic Properties of Triangles

The sum of the angles in a triangle is 180°. This is called the angle-sum property. The sum of the lengths of any two sides of a triangle is greater than the length of the third side. Similarly, the difference between the lengths of any two sides of a triangle is less than the length of the third side. The side opposite to the largest angle is the longest side of the triangle and the side opposite to the smallest angle is the shortest side of the triangle.

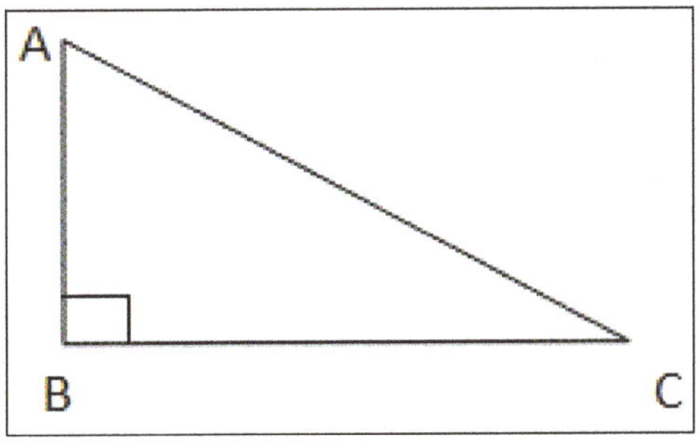

In the figure above, ∠B is the largest angle and the side opposite to it (hypotenuse), is the largest side of the triangle.

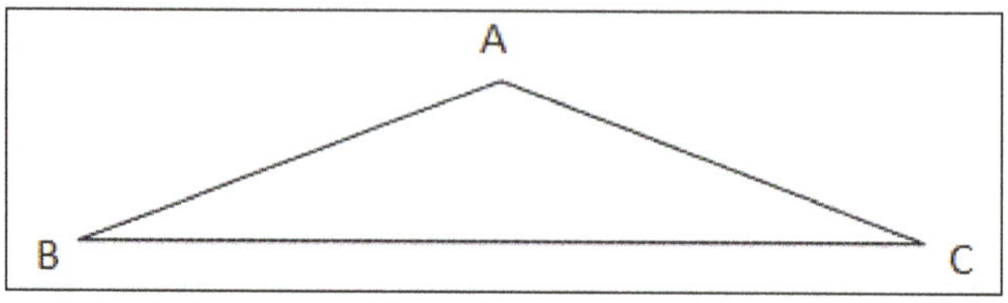

In the figure above, ∠A is the largest angle and the side opposite to it, BC is the largest side of the triangle.

An exterior angle of a triangle is equal to the sum of its interior opposite angles. This is called the exterior angle property of a triangle.

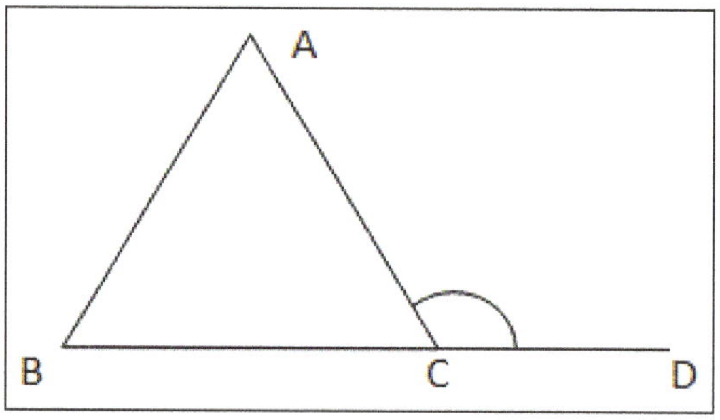

Here, ∠ACD is the exterior angle to the ΔABC. According to the exterior angle property, ∠ACD = ∠CAB + ∠ABC.

Similarity and Congruency in Triangles

Figures with same size and shape are congruent figures. If two shapes are congruent, they remain congruent even if they are moved or rotated. The shapes would also remain congruent if we reflect the shapes by producing mirror images. Two geometrical shapes are congruent if they cover each other exactly. Figures with same shape but with proportional sizes are similar figures. They remain similar even if they are moved or rotated.

Similarity of Triangles

Two triangles are said to be similar if the corresponding angles of two triangles are congruent and lengths of corresponding sides are proportional.

It is written as Δ ABC ~ Δ XYZ and said as Δ ABC 'is similar to' Δ XYZ.

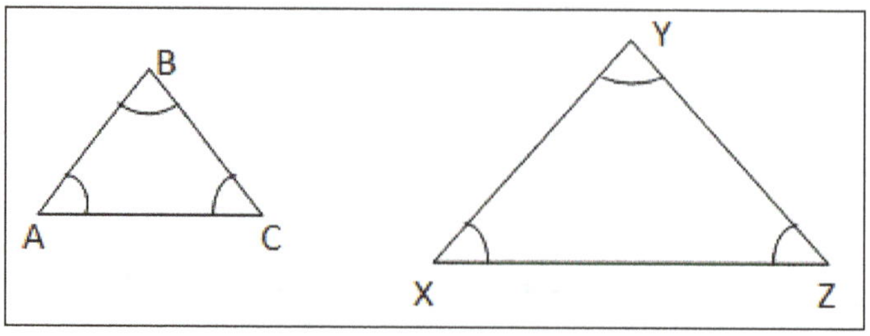

Here, ∠A =∠X, ∠B =∠Y and ∠C = ∠Z and,

AB/XY =BC/YZ = CA/ZX.

The necessary and sufficient conditions for two triangles to be similar are as follows:

Side-Side-Side (SSS) Criterion for Similarity

If three sides of a triangle are proportional to the corresponding three sides of another triangle then the triangles are said to be similar.

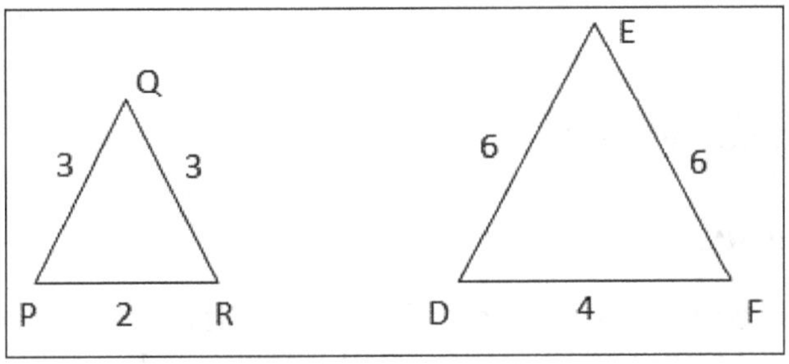

Here, Δ PQR ~ Δ DEF as,

PQ / DE =QR / EF = RP / FD.

Side-Angle-Side (SAS) Criterion for Similarity

If the corresponding two sides of the two triangles are proportional and one included angle is equal to the corresponding included angle of another triangle then the triangles are similar.

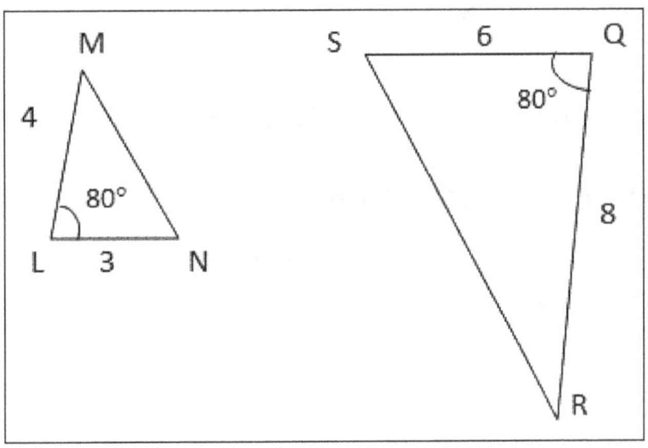

Here, Δ LMN ~ Δ QRS in which,

∠L = ∠Q,

QS/LN = QR/LM.

Angle-Angle-Angle (AAA) Criterion for Similarity

If the three corresponding angles of the two triangles are equal then the two triangles are similar.

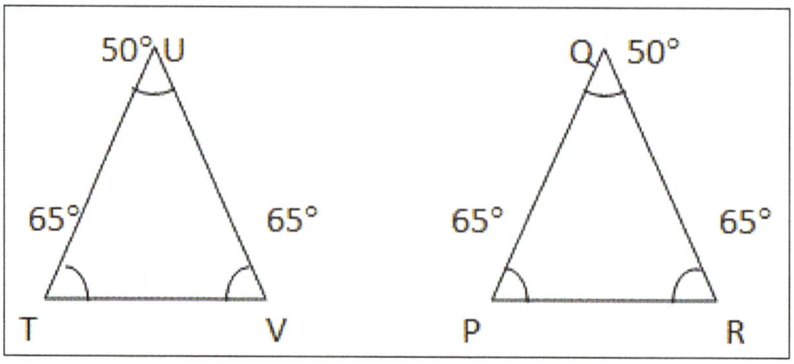

Here Δ TUV ~ Δ PQR as,

∠T = ∠P, ∠U =∠Q and ∠V = ∠R.

Congruency of Triangles

Two triangles are said to be congruent if all the sides of one triangle are equal to the corresponding sides of another triangle and the corresponding angles are equal.

It is written as Δ ABC ≅ Δ XYZ and said as Δ ABC 'is congruent to' Δ XYZ.

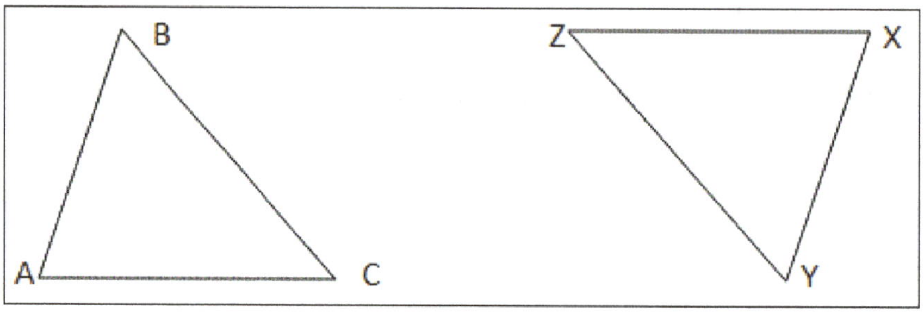

The necessary and sufficient conditions for two triangles to be congruent are as follows:

Side-Side-Side (SSS) Criterion for Congruence

If three sides of a triangle are equal to the corresponding three sides of another triangle then the triangles are said to be congruent.

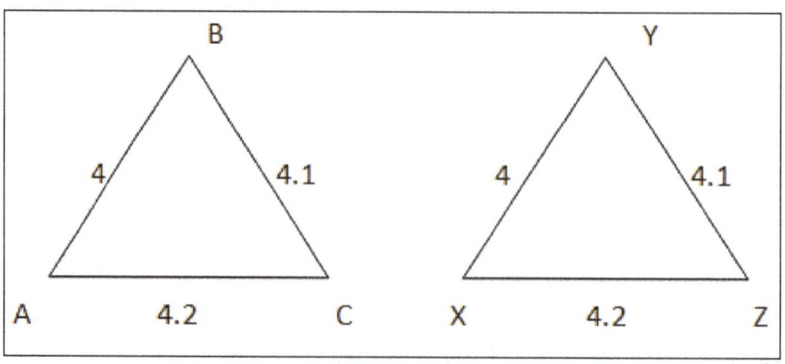

Here, Δ ABC ≅ Δ XYZ as AB = XY, BC = YZ and AC = XZ.

Side-Angle-Side (SAS) Criterion for Congruence

If two sides and the angle included between the two sides of a triangle are equal to the corresponding two sides and the included angle of another triangle, then the triangles are congruent.

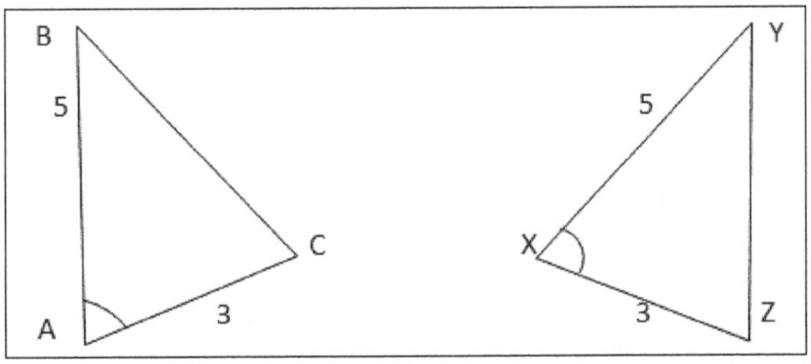

Here, Δ ABC ≅ Δ XYZ as AB = XY, ∠A = ∠X and AC = XZ.

Angle-Side-Angle (ASA) Criterion for Congruence

If two angles and the included side of a triangle are equal to the corresponding two angles and the included side of another triangle then the triangles are congruent.

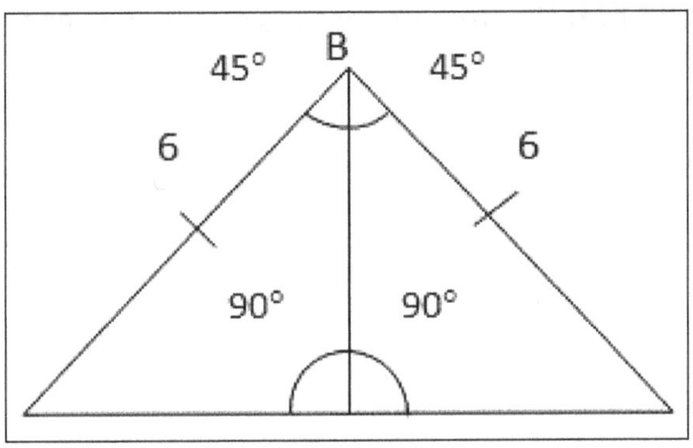

In the figure above, Δ ABD ≅ Δ CBD in which ∠ABD = ∠CBD, AB = CB and ∠ADB = ∠CDB.

Right-Angle Hypotenuse Criterion of Congruence

If the hypotenuse and one side of a right-angled triangle are equal to the corresponding hypotenuse and side of another right-angled triangle, then the triangles are congruent.

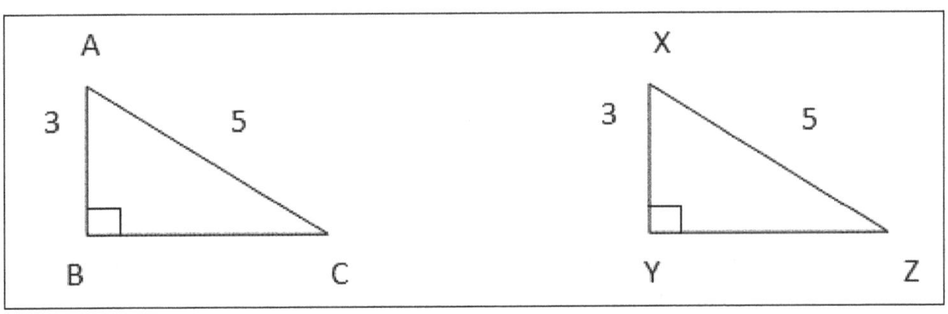

Here, ∠B = ∠Y = 90° and AB = XY, AC = XZ.

Area of a Triangle

The Area of a triangle is given by the formula,

Area of a triangle =(1/2) *Base * Height

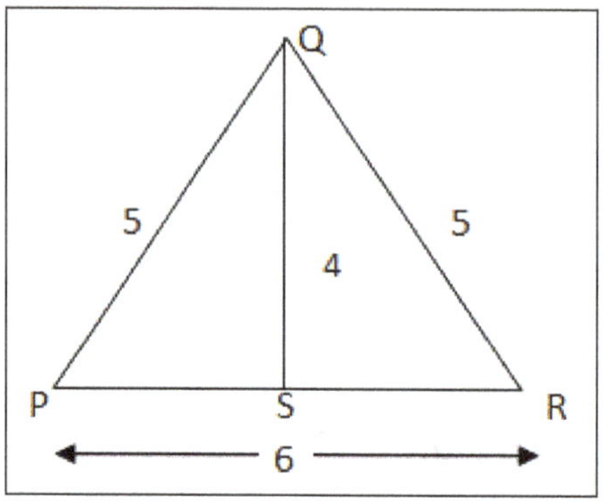

To find the area of a triangle, we draw a perpendicular line from the base to the opposite vertex which gives the height of the triangle. So the area of the Δ PQR = (1/2) * (PR * QS) = (1/2) * 6 *4 =12 sq. units.

For a right triangle, it's easy to find the area as there is a side perpendicular to the base, so we can consider it as height.

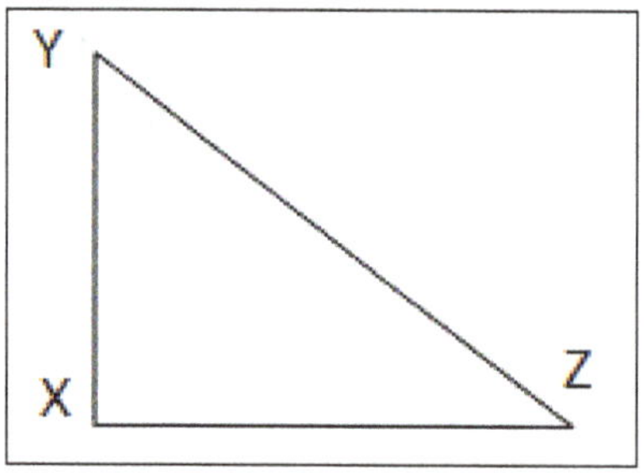

The height of the Δ XYZ is XY and its area is (1/2) * XZ * XY sq. units.

Now, how do we find the area of aWn obtuse triangle LMN?

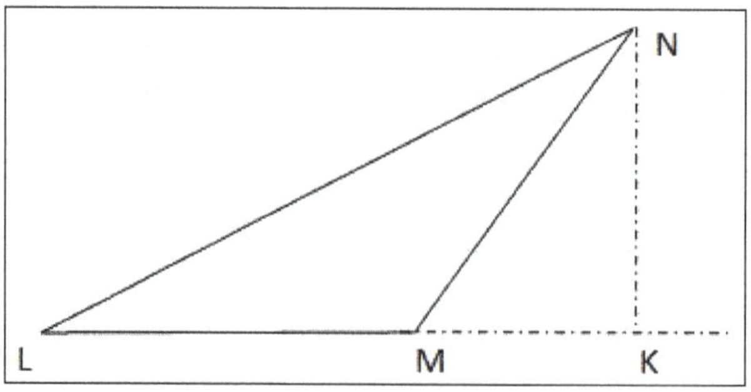

For an obtuse triangle, we extend the base and draw a line perpendicular from the vertex to the extended base which becomes the height of the triangle.

Hence, the area of the Δ LMN = (1/2) * LM * NK sq. units.

Solve the Following

1) Δ ABC is a right triangle and CD ⊥ AB (⊥ stands for 'perpendicular').

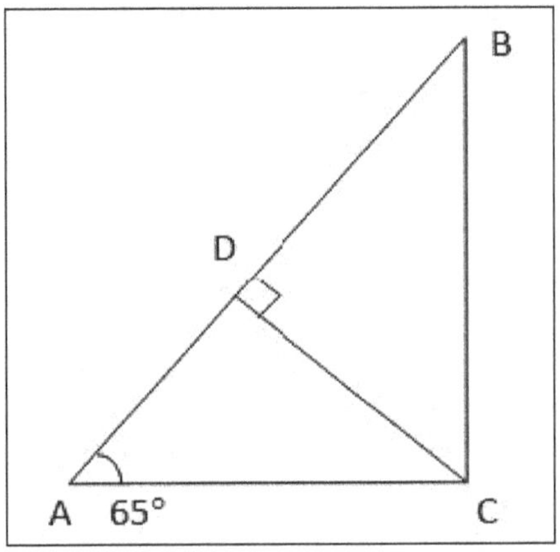

Find i) ∠ACD and ii) ∠ABC.

a. 25, 35

b. 35, 35

c. 25, 25

d. 35, 25

Answer: C.

Explanation: Consider Δ ACD.

∠ADC + ∠DAC + ∠ACD = 180° (since sum of angles in a triangle is 180°),

90 + 65 + ∠ACD = 180° → ∠ACD = 25°,

∠ACD + ∠DCB = 90° → 25 + ∠DCB = 90 → ∠DCB = 65°,

In Δ BCD, ∠DCB + ∠CBD + ∠BDC = 180° (again, sum of all angles in a triangle),

65 + ∠CBD + 90 = 180 → ∠CBD = 25° = ∠ABC.

2) Determine if the following are right triangles:

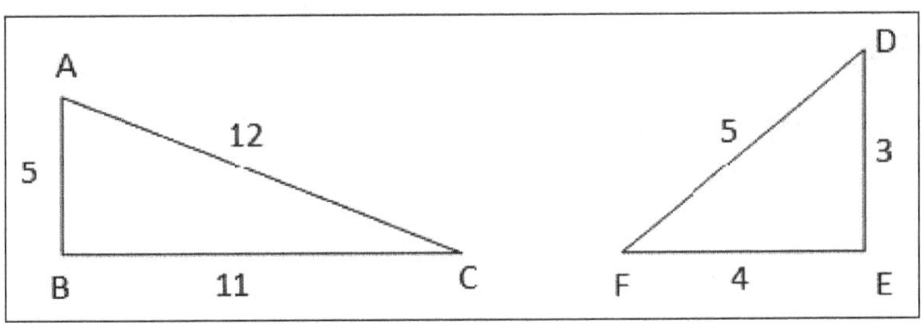

a. Both are right triangles.

b. Δ ABC is not a right triangle, Δ DEF is a right triangle.

c. Δ ABC is a right triangle, Δ DEF is not a right triangle.

d. Both are not right triangles.

Answer: B.

Explanation: The triplet that satisfies the Pythagorean theorem is the set of sides that makes a right triangle.

3) If Δ ABC = 3 (Δ DEF), which of the following is correct?

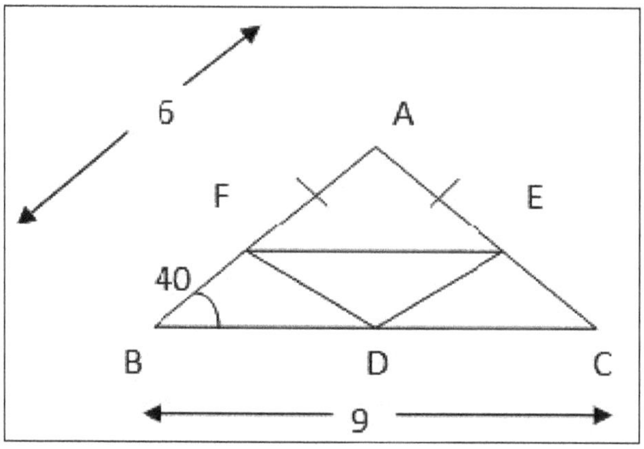

a. ∠E = ∠F = 40°, ∠D = 120° AND DE = DF = 2 and EF = 3.

b. ∠E = ∠F = 40°, ∠D = 110° AND DE = DF = 2 and EF = 3.

c. ∠E = ∠F = 40°, ∠D = 100° AND DE = DF = 2 and EF = 3.

d. ∠E = ∠F = 40°, ∠D = 110° AND DE = DF = 3 and EF = 3.

Answer: C.

Explanation: AB and AC are equal → angles opposite are equal.

Therefore, ∠B =∠C = 40° → ∠A = 100°.

Δ ABC = 3 (Δ DEF) → Δ ABC and Δ DEF are similar.

When two triangles are similar, their corresponding angles are equal and the corresponding sides are proportional.

→ DE =DF = 6/3 = 2 and EF = 3.

→ ∠E =∠F = 40° and ∠D = 100°.

Heron's Formula

Heron's formula is used to find the area of a triangle when we know the length of all its sides. It is also termed as Hero's Formula. We don't have to need to know the angle measurement of a triangle to calculate its area.

$$\text{Area of triangle using three sides} = \sqrt{s(s-a)(s-b)(s-c)}$$

Semiperimeter, s = Perimeter of triangle/2 = (a + b + c)/2.

Hero of Alexandria was a great mathematician who derived the formula for the calculation of the area of a triangle using the length of all three sides. It is also termed as Hero's Formula. He also extended this idea to find the area of quadrilateral and also higher-order polygons. This formula has its huge applications in trigonometry such as proving the law of cosines or law of cotangents, etc.

Heron's Formula for Area of Triangle

According to Heron, we can find the area of any given triangle, whether it is a scalene, isosceles or equilateral, by using the formula, provided the sides of the triangle.

Suppose, a triangle ABC, whose sides are a, b and c, respectively. Thus, the area of a triangle can be given by;

$$\text{Area} = \sqrt{s(s-a)(s-b)(s-c)}$$

Where "s" is semi-perimeter = (a + b + c) / 2

And a, b, c are the three sides of the triangle.

To find the area of a triangle using Heron's formula, we have to follow two steps:

- The first step is to find the value of the semi-perimeter of the given triangle.

  S = (a + b + c)/2

- The second step is to use Heron's formula to find the area of a triangle.

Example: A triangle PQR has sides 4 cm, 13 cm and 15 cm. Find the area of the triangle.

Semiperimeter of triangle PQR, s = (4 + 13 + 15)/2 = 32/2 = 16

By heron's formula, we know;

$$A = \sqrt{[s(s-a)(s-b)(s-c)]}$$

Hence, $A = \sqrt{[16(16-4)(16-13)(16-15)]} = \sqrt{(16 \text{ x } 12 \text{ x } 3 \text{ x } 1)} = \sqrt{576} = 24$ sq.cm

This formula is applicable to all types of triangles. Now let us derive the area formula given by Heron.

Heron's Formula for Quadrilateral

Let us learn how to find the area of quadrilateral using Heron's formula here. If ABCD is a quadrilateral, where AB||CD and AC & BD are the diagonals. AC divides the quad.ABCD into two triangles ADC and ABC. Now we have two triangles here,

Area of quad. ABCD = Area of ΔADC + Area of ΔABC

So, if we know the lengths of all sides of a quadrilateral and length of diagonal AC, then we can use Heron's formula to find the total area. Hence, we will first find the area of ΔADC and area of ΔABC using Heron's formula and at last, will add them to get the final value.

Heron's Formula for Equilateral Triangle

As we know the equilateral triangle has all its sides equal. To find the area of equilateral triangle let us first find the semi perimeter of the equilateral triangle will be:

$$s = (a + a + a)/2$$
$$s = 3a/2$$

where a is the length of the side.

Now, as per the heron's formula, we know,

$$\text{Area} = \sqrt{s(s-a)(s-b)(s-c)}$$

Since, a = b = c

Therefore,

$$A = \sqrt{\left[s(s-a)^3\right]}$$

which is the required formula.

Heron's Formula for Isosceles Triangle

An isosceles triangle has two of its sides equal and the angles corresponding to these sides are congruent. To find the area of isosceles triangle, we can derive the heron's formula as given below:

Let 'a' be the length of the congruent sides and 'b' be the length of the base.

Semi-perimeter (s) =(a + a + b)/2

$$s = (2a + b)/2$$

Using the heron's formula of a triangle,

$$\text{Area} = \sqrt{[s(s-a)(s-b)(s-c)]}$$

By substituting the sides of an isosceles triangle,

$$\text{Area} = \sqrt{[s(s-a)(s-a)(s-b)]}$$
$$= \sqrt{\left[s(s-a)^2(s-b)\right]}$$

Or,

$$= (s-a)\sqrt{[s(s-b)]}$$

which is the required formula to find the area for the given isosceles triangle.

Proof

There are two methods by which we can derive the Hero's formula:

- First, by using trigonometric identities and cosine rule.
- Secondly, solving algebraic expression using Pythagoras theorem.

Proof using Trigonometric Cosine Rule

Let us prove the result using the law of cosines: Let a, b, c be the sides of the triangle and α, β, γ are opposite angles to the sides.

We know that, the law of cosines is:

$$\text{Cos}\gamma = \frac{a^2+b^2-c^2}{2ab}$$

Again, using trig identity, we have,

$$\text{Sin}\gamma = \sqrt{1-\cos^2\gamma}$$

$$= \frac{\sqrt{4a^2b^2-\left(a^2+b^2-c^2\right)^2}}{2ab}$$

Here, Base of triangle = a,

Altitude = b sin γ

Now,

$$A = \frac{1}{2}(\text{base})(\text{altitude}) \quad = \frac{1}{2}ab\sin\gamma$$

$$= \frac{1}{4}\sqrt{4a^2b^2-\left(a^2+b^2-c^2\right)^2}$$

$$= \frac{1}{4}\sqrt{\left(2ab-\left(a^2+b^2-c^2\right)^2\right)\left(2ab+\left(a^2+b^2-c^2\right)\right)}$$

$$= \frac{1}{4}\sqrt{\left(c^2-(a-b)^2\right)\left((a+b)^2-c^2\right)}$$

$$= \sqrt{\frac{(c-(a-b))(c+(a-b))((a+b)-c)((a+b)+c)}{16}}$$

$$= \sqrt{\frac{(b+c-a)}{2}\frac{(a+c-b)}{2}\frac{(a+b-c)}{2}\frac{(a+b+c)}{2}}$$

$$= \sqrt{\frac{(a+b+c)}{2}\frac{(b+c-a)}{2}\frac{(a+c-d)}{2}\frac{(a+b-c)}{2}}$$

$$= \sqrt{s(s-a)(s-b)(s-c)}.$$

Proof using Pythagoras Theorem

Area of a Triangle with 3 Sides.

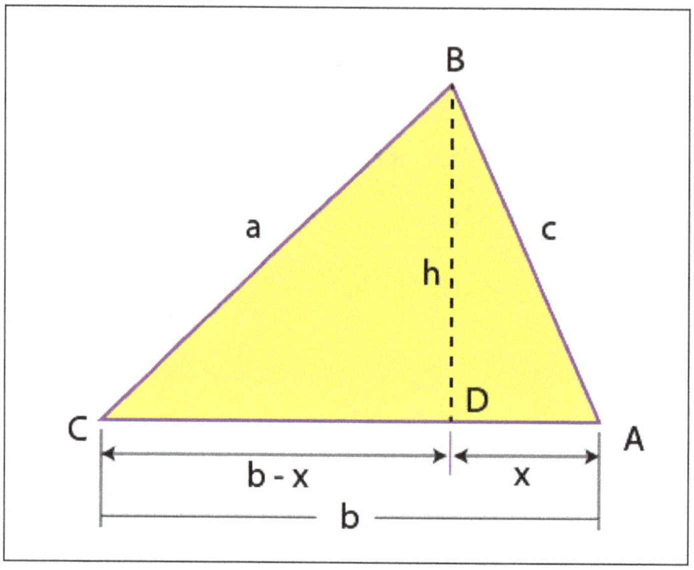

Area of ΔABC is given by, $A = 1/2\ bh$

Draw a perpendicular BD on AC.

Consider a ΔADB,

$$x^2 + h^2 = c^2$$
$$x^2 = c^2 - h^2$$

$$\Rightarrow x = \sqrt{(c^2 - h^2)}$$

Consider a ΔCDB,

$$(b-x)^2 + h^2 = a^2$$
$$(b-x)^2 = a^2 - h^2$$
$$b^2 - 2bx + x^2 = a^2 - h^2$$

Substituting the value of x and x² from equation $x^2 = c^2 - h^2$ and $\Rightarrow x = \sqrt{(c^2 - h^2)}$ we get,

$$b^2 - 2b\sqrt{(c^2 - h^2)} + c^2 - h^2 = a^2 - h^2$$
$$b^2 + c^2 - a^2 = 2b\sqrt{(c^2 - h^2)}$$

Squaring on both sides, we get,

$$(b^2 + c^2 - a^2)^2 = 4b^2(c^2 - h^2)$$

$$\frac{\left(b^2+c^2-a^2\right)^2}{4b^2}=c^2-h^2$$

$$h^2=c^2-\frac{\left(b^2+c^2-a^2\right)^2}{4b^2}$$

$$h^2=\frac{4b^2c^2-\left(b^2+c^2-a^2\right)^2}{4b^2}$$

$$h^2=\frac{(2bc)^2-\left(b^2+c^2-a^2\right)^2}{4b^2}$$

$$h^2=\frac{\left[2bc+\left(b^2+c^2-a^2\right)\right]\left[2bc-\left(b^2+c^2-a^2\right)\right]}{4b^2}$$

$$h^2=\frac{\left[\left(b^2+2bc+c^2\right)-a^2\right]\left[a^2-\left(b^2-2bc+c^2\right)\right]}{4b^2}$$

$$h^2=\frac{\left[(b+c)^2-a^2\right].\left[a^2-(b-c)^2\right]}{4b^2}$$

$$h^2=\frac{\left[(b+c)+a\right]\left[(b+c)-a\right].\left[a+(b-c)\right]\left[a-(b-c)\right]}{4b^2}$$

$$h^2=\frac{(a+b+c)(b+c-a)(a+c-b)(a+b-c)}{4b^2}$$

The perimeter of a ΔABC is,

$$P=a+b+c$$

$$\Rightarrow h^2=\frac{P(P-2a)(P-2b)(P-2c)}{4b^2}$$

$$\Rightarrow h=\sqrt{P(P-2a)(P-2b)(P-2c)}2b$$

Substituting the value of h in equation $A=1/2\ bh$, we get,

$$A=\frac{1}{2}b\frac{\sqrt{P(P-2a)(P-2b)(P-2c)}}{2b}$$

$$A=\frac{1}{4}\sqrt{(P(P-2a)(P-2b)(P-2c))}$$

$$A=\sqrt{\frac{1}{16}P(P-2a)(P-2b)(P-2c)}$$

$$A=\sqrt{\frac{P}{2}\left(\frac{P-2a}{2}\right)\left(\frac{P-2b}{2}\right)\left(\frac{P-2c}{2}\right)}$$

$$\text{Semi perimeter}(s)=\frac{\text{perimeter}}{2}=\frac{P}{2}$$

$$\Rightarrow A=\sqrt{s(s-a)(s-b)(s-c)}$$

Heron's formula is applicable to all types of triangles and the formula can also be derived using the law of cosines and the law of Cotangents.

Problems and Solutions

Example: Find the area of a trapezium, length of whose parallel sides is given as 22 cm and 12 cm and the length of other sides are 14 cm each.

Solution: Let PQRS be the given trapezium in which PQ = 22 cm, SR = 12 cm, PS = QR = 14cm.

Constructions: Draw OR||PS,

Now, PORS is a parallelogram in which PS||OR and PO||SR.

Therefore, PO = SR = 12cm

$\Rightarrow$ OQ = PQ-PO = 22 - 12 = 10cm

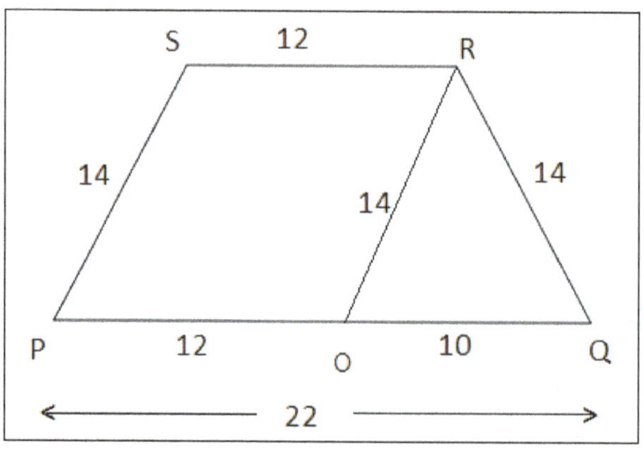

In ΔOQR, we have,

$$s=\frac{14+14+10}{2}=\frac{38}{2}=19$$

Area of $\Delta OQR = \sqrt{s(s-a)(s-b)(s-C)}$

$$= \sqrt{19(19-14)(19-14)(19-10)}$$
$$= \sqrt{4275}$$
$$= 15\sqrt{19}\ cm^2$$

We know that $\text{Area} = \frac{1}{2} \times b \times h,$

$$\Rightarrow 15\sqrt{19} = \frac{1}{2} \times 10 \times h$$
$$\Rightarrow h = 5\sqrt{19}$$

Area of trapezium $\frac{1}{2}(PQ + SR) \times h$

$$= \frac{1}{2}(22 + 12) \times 3\sqrt{19} = 51\sqrt{19} cm^2$$

Example: Find the area of the triangle whose sides measure 10 cm, 17 cm and 21 cm. Also, determine the length of the altitude on the side which measures 17 cm.

Solution: $s = \frac{a+b+c}{2} = \frac{10+17+21}{2} = 24$

Area of Triangle $= \sqrt{s(s-a)(s-b)(s-c)}$

$$\sqrt{24 \times 14 \times 7 \times 3}$$
$$\sqrt{7056} = 84 \text{ square cm}$$

Taking 17 cm as the base length we need to find the height,

Area, A =1/2 x base x height

$1/2 \times 17 \times h = 84$ or $h = 168/17 = 9.88$ cm (Rounded to the nearest hundredth).

Triangles Based on Sides

According to the sides of the triangle, the triangle can be classified into three types, namely:

- Scalene Triangle: A triangle with no equal sides or a triangle in which all the sides are of different length.
- Isosceles Triangle: A triangle with two equal sides and two equal angles is called an isosceles triangle.
- Equilateral Triangle: A triangle in which all three sides are equal, and each interior angle of a triangle measure 60 degrees is called the equilateral triangle.

Equilateral Triangle

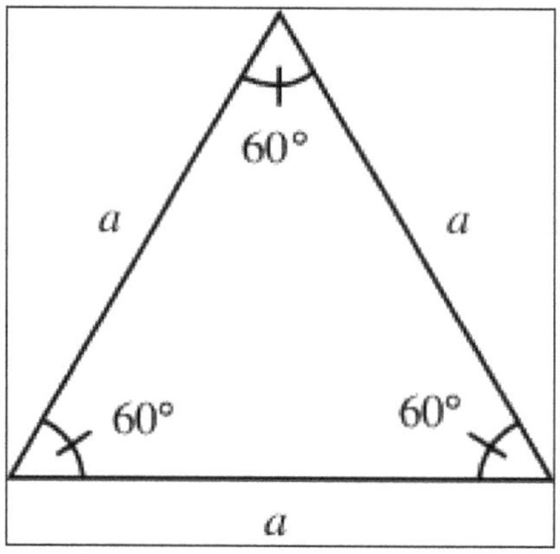

An equilateral triangle is a triangle with all three sides of equal length a, corresponding to what could also be known as a "regular" triangle. An equilateral triangle is therefore a special case of an isosceles triangle having not just two, but all three sides equal. An equilateral triangle also has three equal 60° angles.

The altitude h of an equilateral triangle is,

$$h = a\sin 60° = \frac{1}{2}\sqrt{3}\,a,$$

where a is the side length, so the area is,

$$A = \frac{1}{2}ah = \frac{1}{4}\sqrt{3}\,a^2.$$

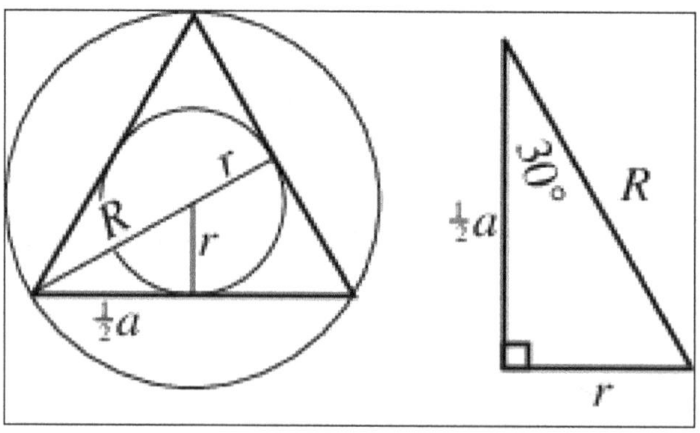

The inradius r, circumradius R, and area A can be computed directly from the formulas for a general regular polygon with side length a and n = 3 sides,

$$r = \frac{1}{2}a\cot\left(\frac{\pi}{3}\right) = \frac{1}{2}a\tan\left(\frac{\pi}{6}\right) = \frac{1}{6}\sqrt{3}\,a$$

$$R = \frac{1}{2}a\csc\left(\frac{\pi}{3}\right) = \frac{1}{2}a\sec\left(\frac{\pi}{6}\right) = \frac{1}{3}\sqrt{3}\,a$$

$$A = \frac{1}{4} n a^2 \cot\left(\frac{\pi}{3}\right) = \frac{1}{4}\sqrt{3}\, a^2.$$

The areas of the incircle and circumcircle are,

$$A_r = \pi r^2 = \frac{1}{2}\pi a^2$$

$$A_R = \pi R^2 = \frac{1}{3}\pi a^2$$

Central triangles that are equilateral include the circumnormal triangle, circumtangential triangle, first Morley triangle, inner Napoleon triangle, outer Napoleon triangle, second Morley triangle, Stammler triangle, and third Morley triangle.

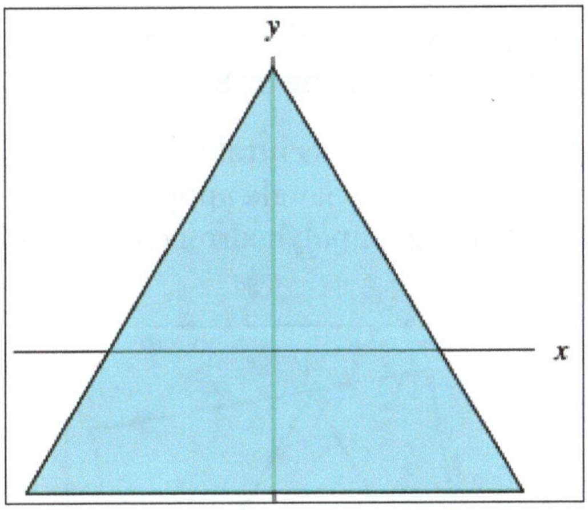

An equation giving an equilateral triangle with R = 1 is given by,

$$\max\left(-2y, y - x\sqrt{3}, y + x\sqrt{3}\right) = 1.$$

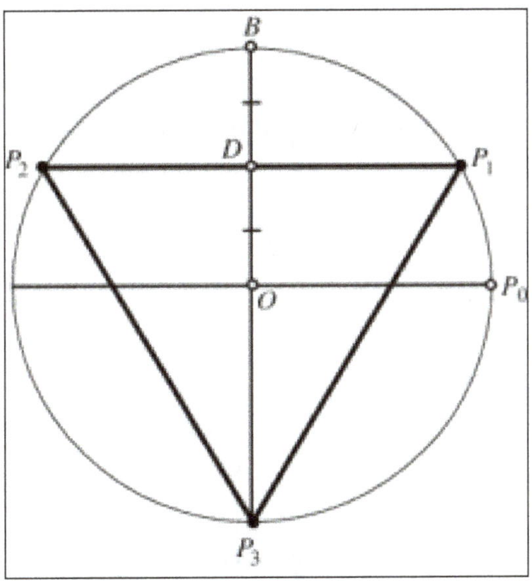

Equilateral triangle construction.

Geometric construction of an equilateral consists of drawing a diameter of a circle OP_0 and then constructing its perpendicular bisector P_3 OB. Bisect OB in point D, and extend the line $P_1 P_2$ through D. The resulting figure $P_1 P_2 P_3$ is then an equilateral triangle. An equilateral triangle may also be constructed from the intersections of the angle trisections of the three interior angles of any triangles (Morley's theorem).

Napoleon's theorem states that if three equilateral triangles are drawn on the legs of any triangle (either all drawn inwards or outwards) and the centers of these triangles are connected, the result is another equilateral triangle. Given the distances of a point from the three corners of an equilateral triangle, a, b, and c, the length of a side s is given by,

$$3\left(a^4+b^4+c^4+s^4\right)=\left(a^2+b^2+c^2+s^2\right)^2$$

There are infinitely many solutions for which a, b, and c are integers. In these cases, one of a, b, c, and s is divisible by 3, one by 5, one by 7, and one by 8.

Begin with an arbitrary triangle and find the excentral triangle. Then find the excentral triangle of that triangle, and so on. Then the resulting triangle approaches an equilateral triangle. The only rational triangle is the equilateral triangle. A polyhedron composed of only equilateral triangles is known as a deltahedron.

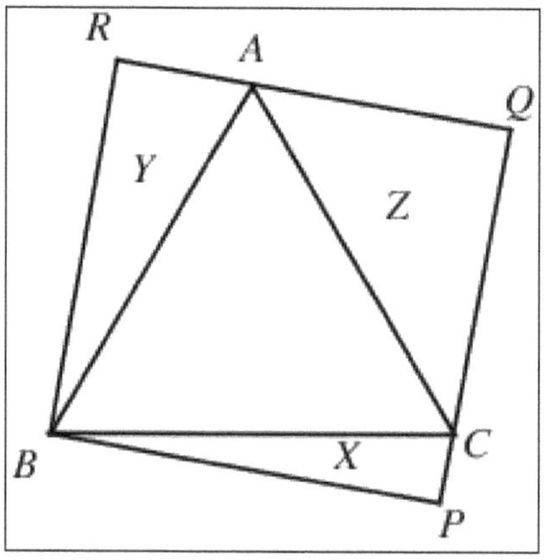

Let any rectangle be circumscribed about an equilateral triangle.

Then, $X+Y=Z$,

where X, Y, and Z are the areas of the triangles in the figure.

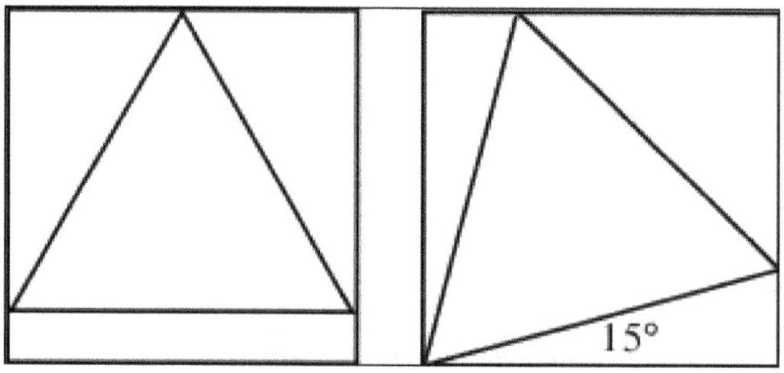

The smallest equilateral triangle which can be inscribed in a unit square (left figure) has side length and area,

$$s = 1$$

$$A = \frac{1}{4}\sqrt{3} \approx 0.4330.$$

The largest equilateral triangle which can be inscribed (right figure) is oriented at an angle of 15° and has side length and area,

$$s = \sec(15^\circ) = \sqrt{6} - \sqrt{2}$$

$$A = 2\sqrt{3} - 3 \approx 0.4641$$

Triangle line picking for points in an equilateral triangle with side lengths a gives a mean line segment length of,

$$\bar{l}_{\Delta(a,a,a)} = \frac{1}{20}(4 + 3\ln 3)a$$
$$= 0.364791843300\ldots a.$$

Isosceles Triangle

An Isosceles triangle is a triangle which has two equal sides. Also, the two angles opposite to the two equal sides are equal. In other words, we can say that "An isosceles triangle is a triangle which has two congruent sides". Suppose in a triangle △ABC, if sides AB and AC are equal, then △ABC is an isosceles triangle where ∠B = ∠C. The theorem that describes the isosceles triangle is "if the two sides of a triangle are congruent, then the angle opposite to them is also congruent".

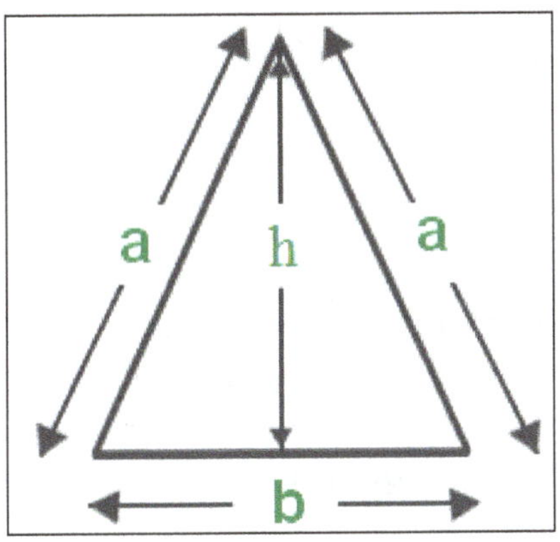

The triangle having all the three sides equal is called an Equilateral triangle. Also, the triangle having all the three unequal sides is called a scalene triangle.

Isosceles Triangle Properties

- As the two sides are equal in this triangle, the unequal side is called the base of the triangle.
- The angles opposite to the two equal sides of the triangle are always equal.

- The altitude of an isosceles triangle is measured from the base to the vertex (topmost) of the triangle.
- A right isosceles triangle has the third angle as 90 degrees.

Generally, the isosceles triangle is classified into different types namely:

- Isosceles acute triangle.
- Isosceles right triangle.
- Isosceles obtuse triangle.

Isosceles Acute Triangle

As we know that the different dimensions of a triangle are legs, base, and height. All the isosceles triangle has an axis of symmetry along the perpendicular bisector of its base. Depends on the angle between the two legs, the isosceles triangle is classified as acute, right and obtuse. The isosceles triangle can be acute if the two angles opposite to the legs are equal and are less than 90 degrees (acute angle).

Isosceles Right Triangle

A right isosceles triangle has two equal sides, wherein one of the two equal sides acts as perpendicular and another one as a base of the triangle. The third side, which is unequal, is termed as the hypotenuse. Therefore, we can apply here the famous Pythagoras theorem, where the square of the hypotenuse is equal to the sum of the square of base and perpendicular. Suppose, the sides of the right isosceles triangle are a, a, and h, where a is the two equal sides and h is the hypotenuse, then,

$$h = \sqrt{(a^2 + a^2)} = \sqrt{2a^2} = a\sqrt{2}$$

$$\text{or } h = \sqrt{2}\, a$$

Isosceles Obtuse Triangle

We know that the obtuse triangle is a triangle in which one of its angles is greater than 90 degrees (right angle). Also, it is not possible to draw a triangle with more than two obtuse angles. We know that the obtuse triangle can be a scalene triangle or isosceles triangle. Therefore, the isosceles obtuse triangle is a triangle, which has two equal sides with an obtuse angle.

Isosceles Triangle Formulas

We know that an isosceles triangle is a two-dimensional shape with three sides. The measures to compute the isosceles triangle are the area and perimeter.

Isosceles Triangle Area

The area of an isosceles triangle is defined as the region occupied by it in the two-dimensional space. Generally, the isosceles triangle is half the product of the base and height of an isosceles triangle. The formula to calculate the area of an isosceles triangle is given by:

The area of an isosceles triangle A = ½ × b × h Square units, where b is the base and h is the height of the triangle.

Isosceles Triangle Perimeter

As we know the perimeter of any shape is the boundary of the shape. Similarly, the perimeter of an isosceles triangle is defined as the sum of the three sides of an isosceles triangle. The perimeter of an isosceles triangle can be found if we know its base and side. The formula to calculate the perimeter of the isosceles triangle is given by:

The perimeter of an isosceles Triangle, P = 2a + b units, where 'a' is the length of the two equal legs of an isosceles triangle and b is the base of the triangle.

Isosceles Triangle Altitude

When an altitude is drawn to the base of the isosceles triangle, it bisects the vertex angle. As it bisects the base, the two congruent triangles are created. The altitude of the triangle forms the required right angle and the altitude becomes the shared legs. Also, the congruent legs of a triangle become the congruent hypotenuse. Therefore, the altitude of drawn to the base of the isosceles triangle bisects the base.

Example: Find the area of an isosceles triangle given its height as 6 cm and base as 4 cm?

Solution: Given that,

Base = 4 cm and height = 6 cm

We know that the area of an isosceles triangle is ½ × b × h square units. Now, substitute the base and height value in the formula,

Area of an isosceles triangle is ½ × b × h

$$A = \frac{1}{2} \times 4 \times 6 = 12\ cm^2$$

Therefore, the area of an isosceles triangle is 12 cm^2.

Example: Find the perimeter of an isosceles triangle, with side 6 cm and base 4 cm.

Solution:

Given: Base = 4 cm,

Length of the two equal arms = 6 cm,

We know that the formula to calculate the perimeter of an isosceles triangle is P = 2a + b units,

Now, substitute the values in the perimeter formula, we get,

$$P = 2(6) + 4 = 12 + 4 = 16\ cm$$

Hence, the perimeter of an isosceles triangle is 16 cm.

Scalene Triangle

Scalene triangles are triangles with each side with a different length. They are unusual in the way, which they are not. Most triangles drawn at random will be of this kind. The interior angles of a scalene triangle are all also different. Its converse is also true.

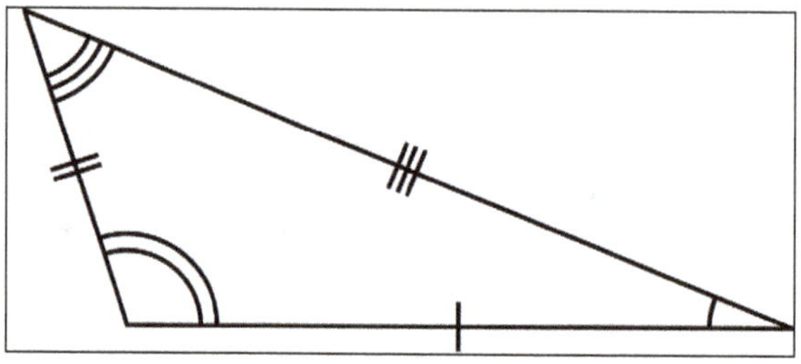

Thus, if a triangle has 3 unequal sides and also the angles of the triangle are different then it will be definitely a scalene triangle. It's some notable properties are the triangle with 3 unequal sides, 3 unequal angle measures and having no line of symmetry. It has no point symmetry at all. The angles in the triangle may be an acute, obtuse or right angle.

Properties of the Scalene Triangle

All angles of a scalene triangle are unequal.	A scalene triangle has no line of symmetry.	The angle opposite to the longest side would be the greatest angle and vice versa.
All sides of the given scalene triangle are unequal. 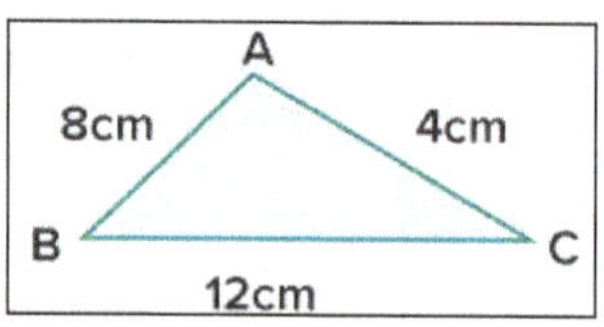	The given triangle cannot be divided into two identical halves. There is no line of symmetry. 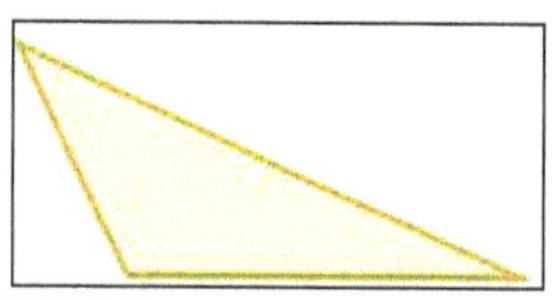	In this triangle, angle E will be the greatest angle as it is opposite to the greatest side GO and angle O will be the smallest angle as it is opposite to the smallest side GE. 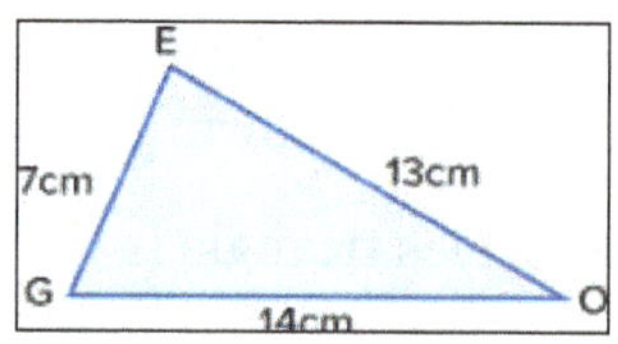

A scalene triangle can be acute-angled or obtuse-angled or right-angled.

Obtuse scalene triangle	Acute scalene triangle	Right scalene triangle
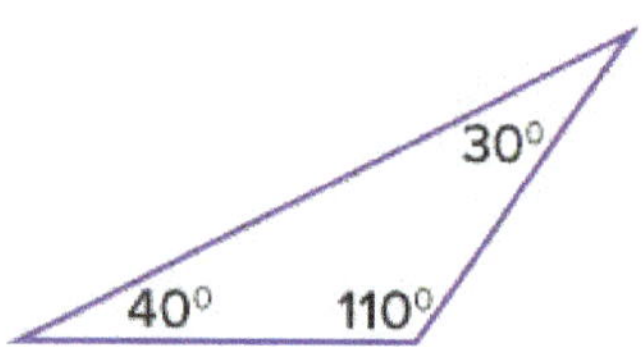	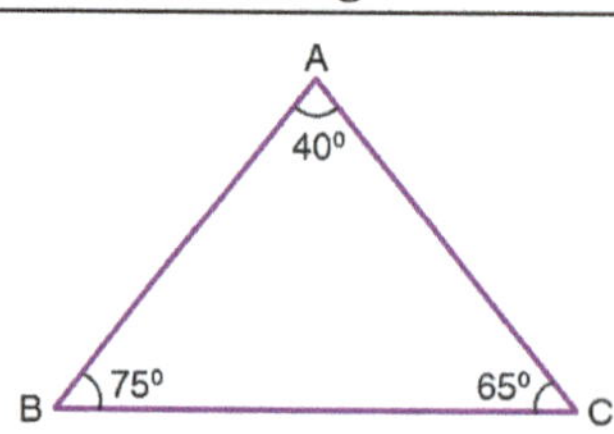	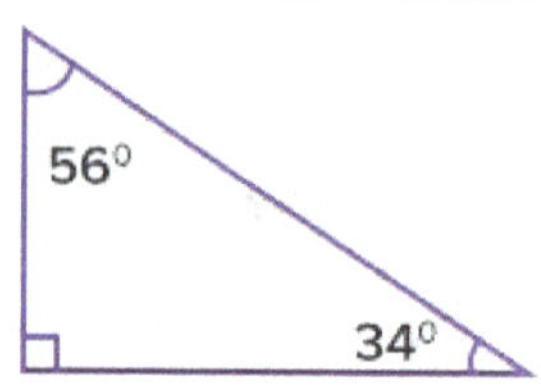

The Formula for Scalene Triangle

Some useful scalene triangle formulas are as follows:

- Area of Triangle $= \frac{1}{2} \times b \times h$, where b is the base and h is the height.

- The formula for Perimeter of any Triangle is, P = a + b + c where a, b, c are the length of the sides.
- When all three sides are given then the area of the triangle will be, $A=\sqrt{[s\times(s-a)\times(s-b)\times(s-c)]}$, where s is the semi-perimeter of the triangle. Also, $o_s=\frac{1}{2}(a+b+c)$.

Solved Examples for Scalene Triangle Formula

Example: If the base of a triangle is 6 and height is 12cm, find the area of this type of scalene triangle.

Solution: Given in the problem,

b = 6 cm

h = 12 cm

thus are of this triangle will be,

$$\text{Area}=\frac{1}{2}\times b\times h$$

$$\text{Area}=\frac{1}{2}\times 6\times 12=36\,\text{cm}^2$$

Example: Find the perimeter of a triangle whose sides are of the lengths 6 cm, 8 cm and 6 cm.

Solution: Given length of sides are,

a = 6 cm

b = 8 cm

c = 6 cm

Therefore, perimeter of this triangle will be,

= a + b + c = 6 + 8 + 6 = 20 cm

Example: Find the area of the scalene triangle ABC with the sides 8cm, 6cm and 4cm.

Solution:

a = 8 cm

b = 6 cm

c = 4 cm

Also, $s=\frac{1}{2}(a+b+c)$

s = 9 cm

Now formula for area is,

$$\text{Area} = \sqrt{[s\times(s-a)\times(s-b)\times(s-c)]}$$

$$\text{Area} = \sqrt{[9\times(9-8)\times(9-6)\times(9-4)]} = \sqrt{135} = 11.6\text{cm}^2$$

Therefore, the area of the scalene triangle will be 11.6 square cm.

Triangles Based on Angles

According to the interior angles of the triangle, it can be classified as three types, namely:

- Acute Angle Triangle: A triangle which consists of three acute angles. It means that all the angles are less than 90 degrees.
- Right Angle Triangle: A triangle in which one angle measures 90 degrees and other two angles are less than 90 degrees (acute angles).
- Obtuse Angle Triangle: A triangle in which one angle measures above 90 degrees and the other two angles measures less than 90 degrees.

Acute Triangle

An acute-angled triangle is a type of triangle in which all the three internal of the triangle are acute, that is, they measure less than 90°. Here are some examples of acute triangles:

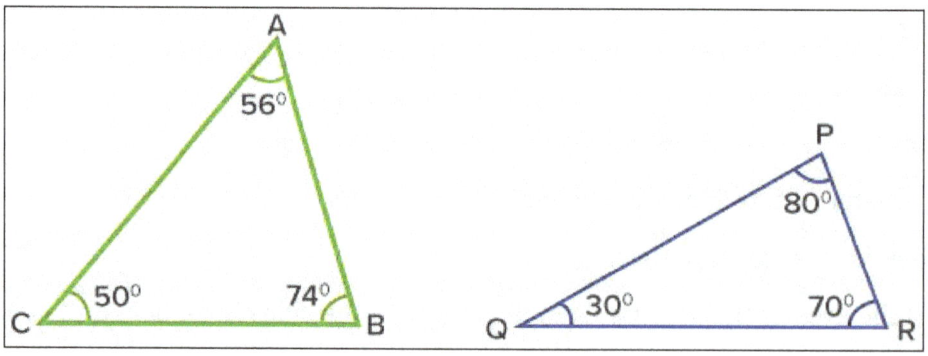

Types of Acute Triangles

There are broadly three types of acute triangles:

Equilateral Acute Triangle	Isosceles Acute Triangle	Scalene Acute Triangle
All the interior angles of an equilateral acute triangle measure 60°. It is also known as equiangular triangle.	Two angles of an isosceles acute triangle that measure the same, just like its two sides.	All three sides and internal angles of a scalene acute triangle are unequal, that is, all angles measure less than 90 degrees.

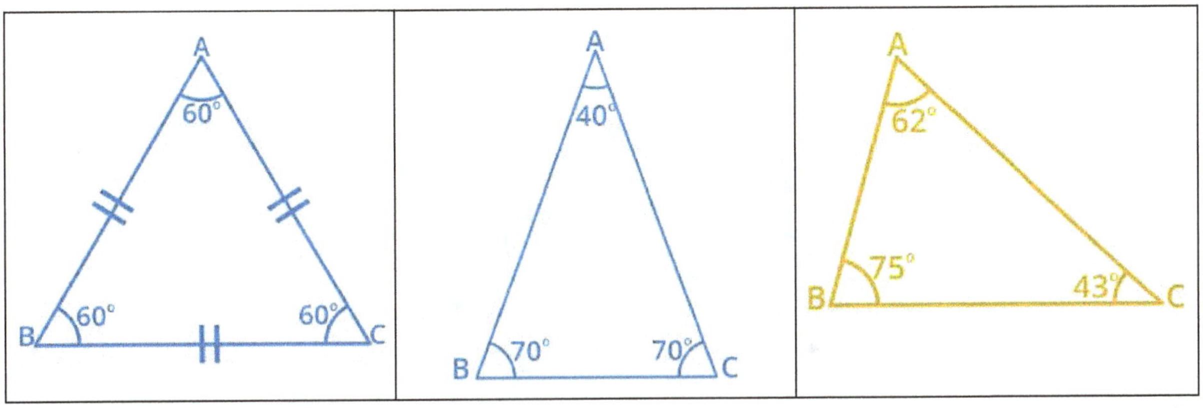

Area of an Acute Triangle

Area of a triangle $= \frac{1}{2}$ Base X Height

Here AD is the height, and BC is the base. So, the area can be calculated by simply putting in the values in the above formula.

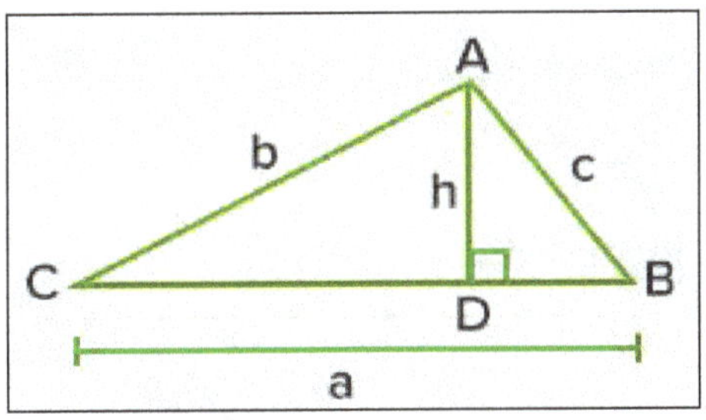

General Properties of Acute Triangle

Given below are a few general properties of acute triangles:

Example: All 3 interior angles of the triangle are acute. You can note that all the angles of the triangle are less than 90°.

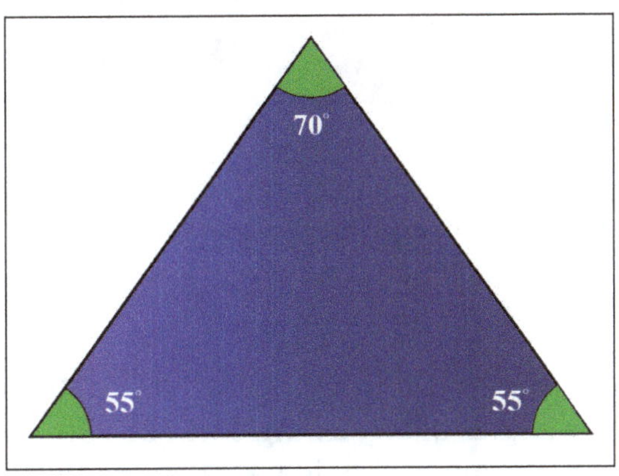

Example:

Area of a triangle is $A = \frac{1}{2} \times b \times h$

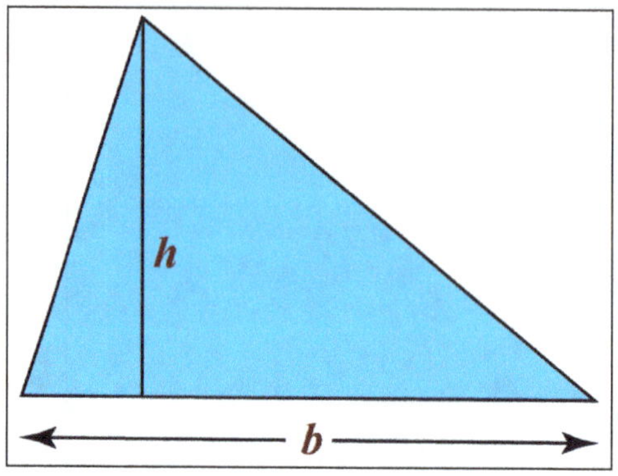

The height will be the perpendicular drawn from the vertex to the base of the triangle. And the area will be calculated from height and base as:

$$A = \frac{1}{2} \times b \times h$$

Example:

The perimeter of the triangle is $P = a + b + c$,

And the semi perimeter is $s = \frac{a+b+c}{2}$

These are the few acute triangle formulas.

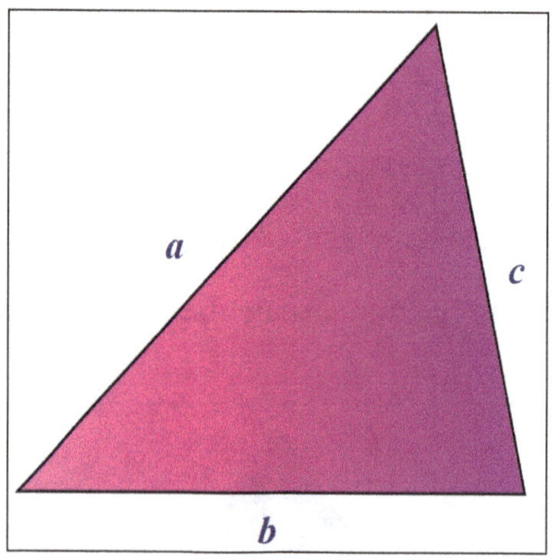

Look at the values of Perimeter and semi - perimeter shown in the diagram given above.

Angles of Acute Triangle

The acute triangle angles can be anything as long as each one of them lies between 0° to 90°. But we also know that the sum of the angles of any triangle will be 180°. So, overall to be considered as an acute triangle, the angles of the triangle should be following the given two rules:

- Each angle lies between 0° to 90°.
- The sum of all 3 angles of the triangle will be 180°.

Try inserting values of any two angles in the boxes given above and see what will be the third angle.

Solved Examples

Example: Which of the following angle measures can form an acute triangle?

a. 60°, 70°, 50°

b. 95°, 30°, 55°

c. 90°, 45°, 45°

d. 90°, 60°, 30°

Solution: An acute triangle has all of the angles < 90°.

Among the given options, option (a) satisfies the condition.

∴ Option "a" forms an acute triangle.

Example: Given below is an isosceles triangle ABC, is this an acute triangle?

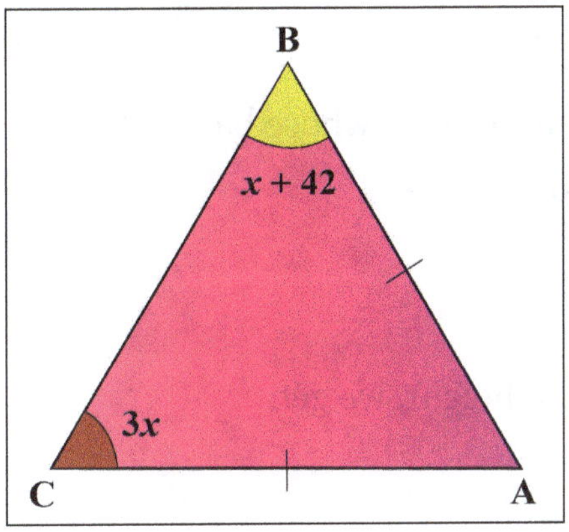

Solution:

$$
\begin{aligned}
3x &= x + 42 (\because \angle ABC = \angle BCA) \\
\therefore 2x &= 42 \\
x &= 21 \\
\angle ABC &= x + 42 \\
&= 63^\circ \\
\Rightarrow \angle BCA &= 63^\circ (\because 3x = 3 \times 21 = 63) \\
\therefore \angle BAC &= (180 - (63 + 63)) \\
&= 180 - 126 \\
&= 54^\circ
\end{aligned}
$$

Hence we finally get: ∠ABC = ∠BCA = 63° and ∠BAC = 54°.

∴ ABC is an acute triangle

Example: A triangle with one exterior angle measuring 80° is shown in the image. Can this triangle be an acute triangle?

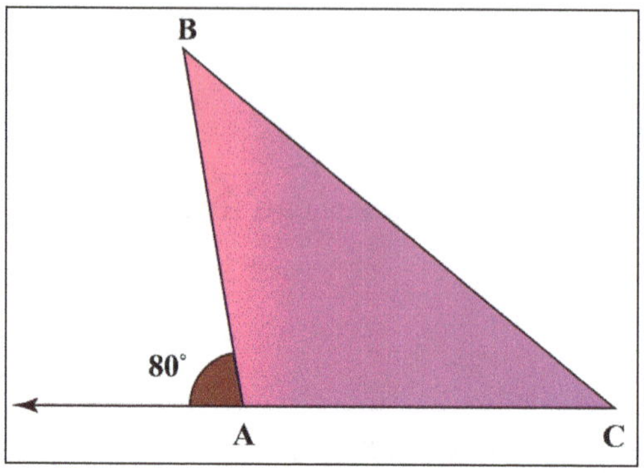

Solution: The exterior angle and the adjacent interior angle form a linear pair (i.e., they add up to 180°).

Hence, the interior angle at vertex B is:

$$180° - 80° = 100°$$

∴ The given triangle is obtuse-angled.

Example: Find the area of an acute triangle whose base is 8 in and height is 4 in.

Solution: Area of an acute-angled triangle,

$$\frac{1}{2} \times \text{base} \times \text{height}$$

Substituting the values of base and height, we get:

$$\text{Area} = \frac{1}{2} \times 8 \times 4 = 16\,\text{in}^2$$

∴ Area of the triangle = 16 in^2

Example: Find the height of an acute triangle whose area = 60 in^2 and base = 8 in.

Solution:

Area of an acute-angled triangle,

$$\frac{1}{2} \times \text{base} \times \text{height}$$

Therefore, height of the acute triangle can be calculated by:

$$\text{Height} = \frac{2 \times \text{Area}}{\text{base}}$$

Substituting the values, we get:

$$\text{Height} = \frac{2 \times 60}{8} = 15\,\text{in}$$

∴ Height = 15in

Right Triangle

A right triangle is triangle with an angle of 90° ($\pi/2$ radians). The sides a, b, and c of such a triangle satisfy the Pythagorean theorem, $a^2 + b^2 = c^2$,

where the largest side is conventionally denoted c and is called the hypotenuse. The other two sides of lengths a and b are called legs, or sometimes catheti.

The favorite A-level math exam question of the protagonist Christopher in the novel The Curious Incident of the Dog in the Night-Time asks for proof that a triangle with sides of the form n^2+1, n^2-1, and 2 n where n > 1 is a right triangle, and that the converse does not hold.

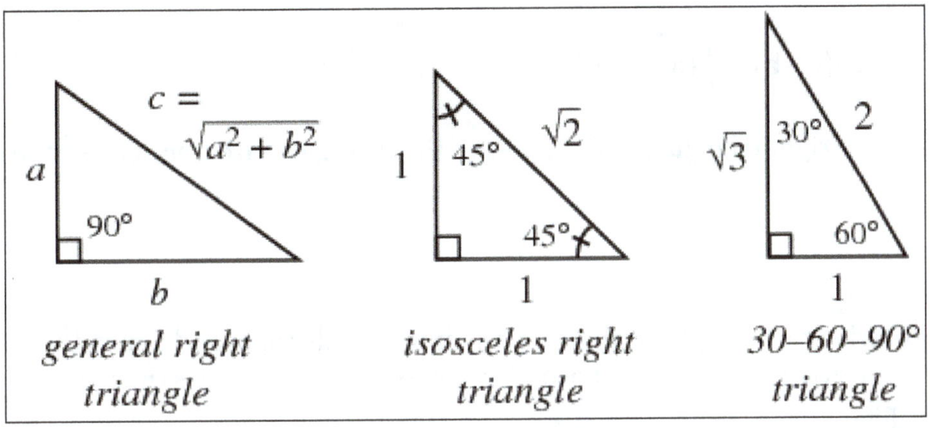

The side lengths (a, b, c) of a right triangle form a so-called Pythagorean triple. A triangle that is not a right triangle is sometimes called an oblique triangle. Special cases of the right triangle include the isosceles right triangle (middle figure) and 30-60-90 triangle (right figure).

For any three similar shapes of area A_i on the sides of a right triangle, $A_1 + A_2 = A_3$,

which is equivalent to the Pythagorean theorem.

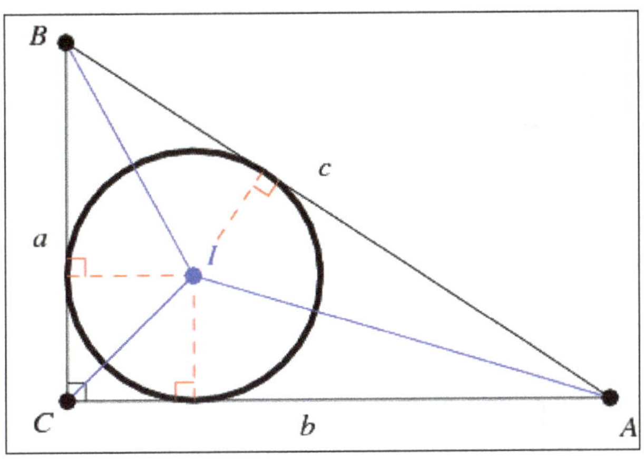

For a right triangle with sides a, b, and hypotenuse c, the area is simply,

$$A = \frac{1}{2}ab.$$

The inradius can be found by equating the area of the triangle ΔABC with the sum of the areas of the three triangles ΔABI, ΔACI, and ΔBCI having the inradii as altitudes, giving,

$$\frac{1}{2}ab = \frac{1}{2}ra + \frac{1}{2}rb + \frac{1}{2}rc = \frac{1}{2}r(a+b+c).$$

Solving for r then gives,

$$r = \frac{ab}{a+b+c}.$$

This can also be written in the equivalent forms,

$$r = \sqrt{\frac{1}{2}(c-a)(c-b)} = \frac{1}{2}(a+b-c).$$

The hypotenuse of a right triangle is a diameter of the triangle's circumcircle, so the circumradius is given by,

$$R = \frac{1}{2}c.$$

A primitive right triangle is a right triangle having integer sides a, b, and c such that GCD (a, b, c) = 1, where GCD is the greatest common divisor. The set of values (a, b, c) is then known as a primitive Pythagorean triple.

For a right triangle with integer side lengths, any primitive Pythagorean triple can be written,

$$a = m^2 - n^2$$
$$b = 2mn$$
$$c = m^2 + n^2.$$

Using these, equation $r = \sqrt{\frac{1}{2}(c-a)(c-b)}$ becomes,

$$r = \frac{(m^2 - n^2)2mn}{m^2 - n^2 + 2mn + m^2 + n^2}$$
$$= n(m-n),$$

which is an integer whenever m and n are integers.

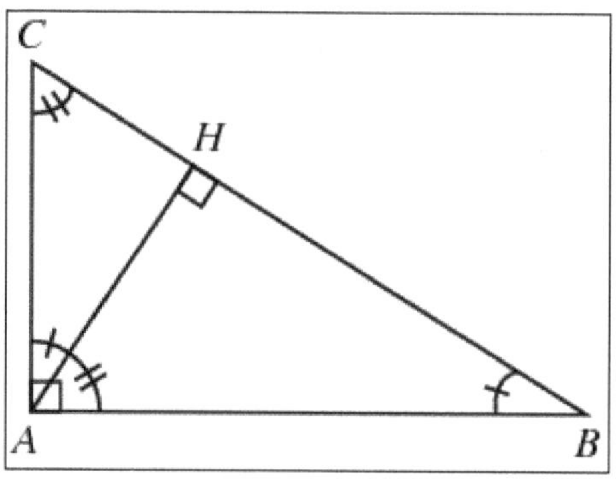

Given a right triangle ΔABC, draw the altitude AH from the right angle A. Then the triangles ΔAHC and ΔBHA are similar.

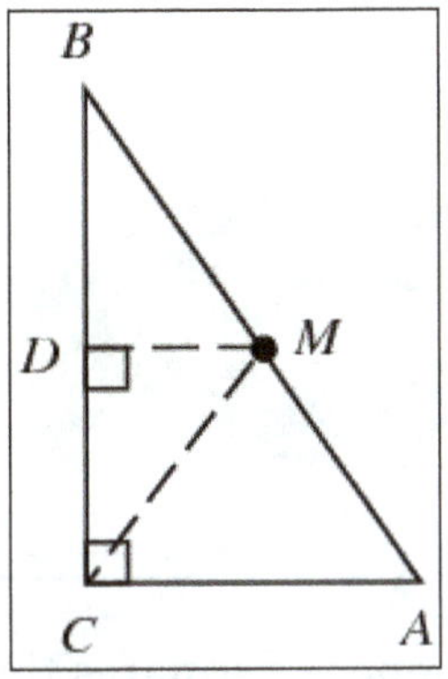

In a right triangle, the midpoint of the hypotenuse is equidistant from the three polygon vertices. This can be proved as follows. Given ΔABC, let M be the midpoint of AB (so that AM = BM). Draw DM||CA, then since ΔBDM is similar to ΔBCA, it follows that BD = DC. Since both ΔBDM and ΔCDM are right triangles and the corresponding legs are equal, the hypotenuses are also equal, so we have AM = BM = CM and the theorem is proved.

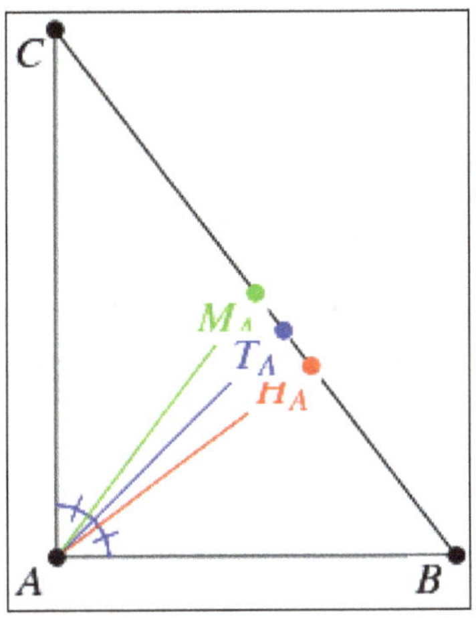

In addition, the triangle median AM_A and altitude AH_A of a triangle ΔABC are reflections about the angle bisector AT_A of A iff ΔABC is a right triangle.

Fermat showed how to construct an arbitrary number of equiareal non-primitive right triangles. An analysis of Pythagorean triples demonstrates that the right triangle generated by a triple $(m_i^2 - n_i^2, 2m_in_i, m_i^2 + n_i^2)$ has common area,

$$A = rs(2r+s)(r+2s)(r+s)(r-s)(r^2+rs+s^2)$$

The only extremum of this function occurs at (r, s) = (0, 0). Since A(r, s) = 0 for r = s, the smallest area shared by three non-primitive right triangles is given by (r, s) = (1, 2), which results in an area of 840 and corresponds to the triplets (24, 70, 74), (40, 42, 58), and (15, 112, 113). One can also find quartets of right triangles with the same area. The quartet having the smallest known area is (111, 6160, 6161), (231, 2960, 2969), (518, 1320, 1418), (280, 2442, 2458), with area.

It is also possible to find sets of three and four right triangles having the same perimeter.

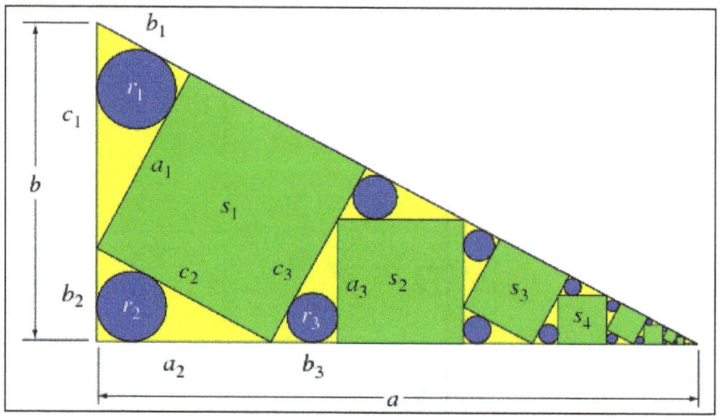

In a given right triangle, an infinite sequence of squares that alternately lie on the hypotenuse and longest leg can be constructed. These create a sequence of increasingly smaller similar right triangles. Let the original triangle have legs of lengths a and b and hypotenuse of length $c = \sqrt{a^2 + b^2}$. Also define,

$$x \equiv \frac{ac}{ab + c^2}$$

$$y \equiv \frac{1}{2}\sqrt{2\left(c^2 - (a+b)c + ab\right)}.$$

Then the sides of the n square are of length, $s_n = bx^n$.

Number the upper left triangle as 1, and then the remainder by following the "strip" of triangles at adjoining vertices. Then the side lengths of these triangles are,

$$a_n = \begin{cases} s_{(n+1)/2} & \text{for n odd} \\ \frac{ab}{c} x^{n/2} & \text{for n even} \end{cases}$$

$$b_n = \begin{cases} \frac{b^2}{a} x^{(n+1)/2} & \text{for n odd} \\ \frac{b^2}{c} x^{n/2} & \text{for n even} \end{cases}$$

$$c_n = \begin{cases} \frac{bc}{a} x^{(n+1)/2} & \text{for n odd} \\ s_{n/2} & \text{for n even.} \end{cases}$$

The inradii of the corresponding triangles can be found from,

$$r_n = \frac{1}{2}\sqrt{\frac{(b_n + c_n - a_n)(c_n + a_n - b_n)(a_n + b_n - c_n)}{a_n + b_n + c_n}},$$

$$\text{giving, } r_n = \begin{cases} \dfrac{b}{a} y\, x^{(n+1)/2} & \text{for n odd} \\ \dfrac{b}{c} y\, x^{n/2} & \text{for n even.} \end{cases}$$

A Sangaku problem from 1913 in the Miyagi Prefecture asks for the relationships between the first, third, and fifth inradii. This can be solved using elementary trigonometry as well as the explicit equations given above, and has solution,

$$r_3 = \sqrt{r_1 r_5}.$$

Obtuse Triangle

An obtuse triangle is a type of triangle where one of the vertex angles is greater than 90°.

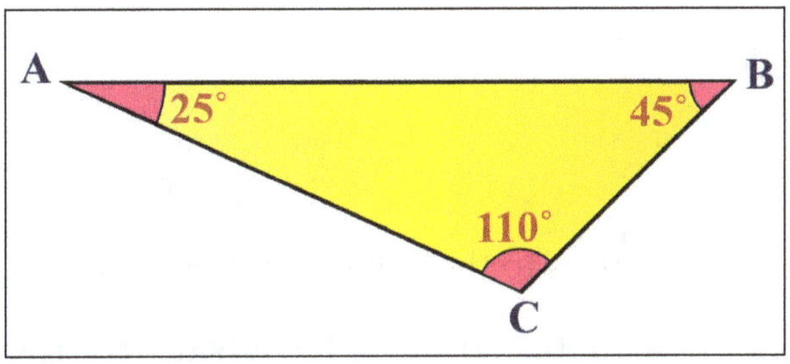

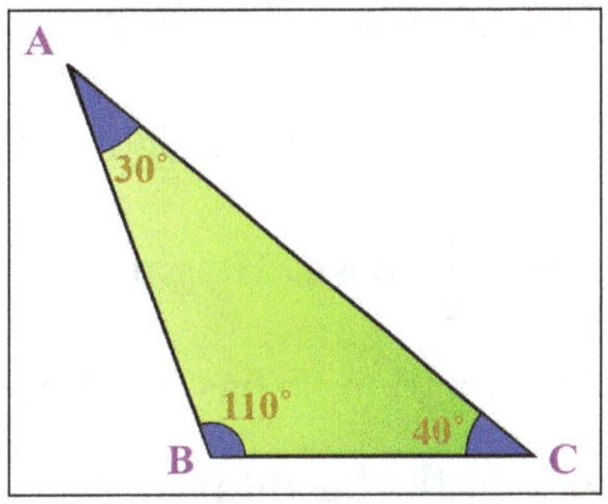

The triangles above have one angle greater than 90°. Hence, they are called obtuse-angled triangle or simply obtuse triangle.

An obtuse-angled triangle can be scalene or isosceles, but never equilateral.

Obtuse Angled Triangle Formula

Perimeter

The perimeter of an obtuse triangle is the sum of the measures of all its sides.

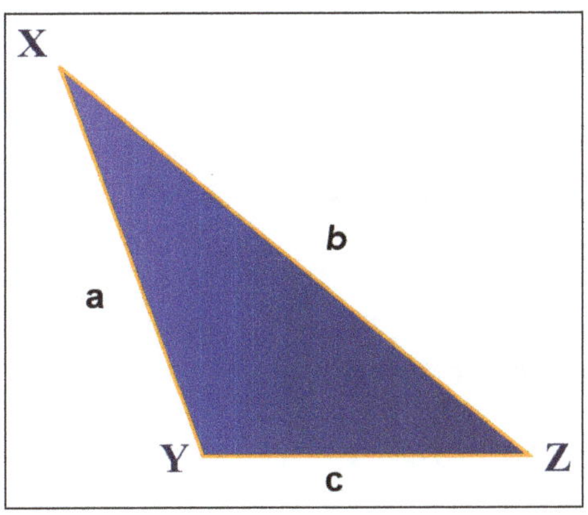

Perimeter of obtuse triangle $= a + b + c$.

Area

In the given obtuse triangle ABC,

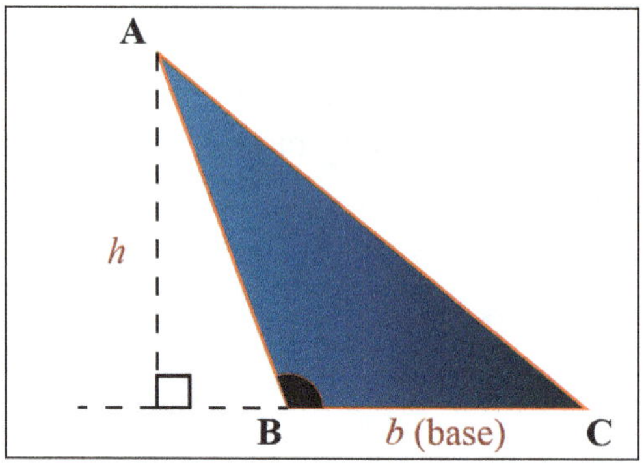

$$\text{Area of } \Delta ABC = \frac{1}{2} h \times b$$

BC is the base and h is the height of the triangle.

$$\text{Area} = \frac{1}{2} \times \text{Base} \times \text{Height}$$

We know that a triangle has 3 altitudes from the 3 vertices to the corresponding opposite sides.

The altitude or the height from the acute angles of an obtuse triangle lie outside the triangle. We extend the base as shown and determine the height of the obtuse triangle. We can also find the area of an obtuse triangle area using Heron's formula.

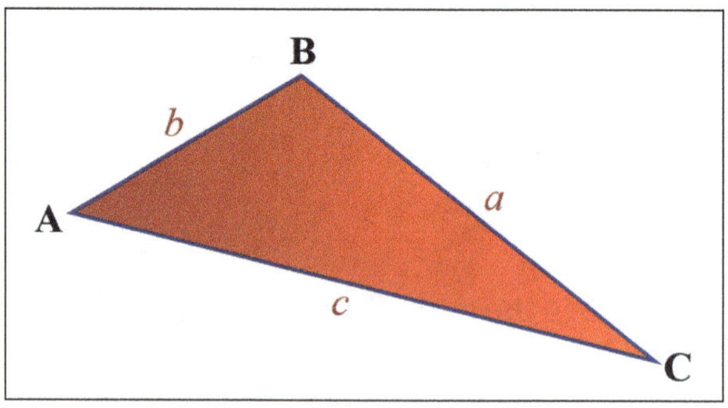

Consider the triangle ABC with sides a, b and c.

Heron's formula to find the area of a triangle is:

$$\sqrt{s(s-a)(s-b)(s-c)}$$

Note that (a + b + c) is the perimeter of the triangle.

s is the semi-perimeter which is given by:

$$s = \frac{a+b+c}{2}.$$

Obtuse Angled Triangle Properties

Example: The longest side of a triangle is the side opposite to the obtuse angle.

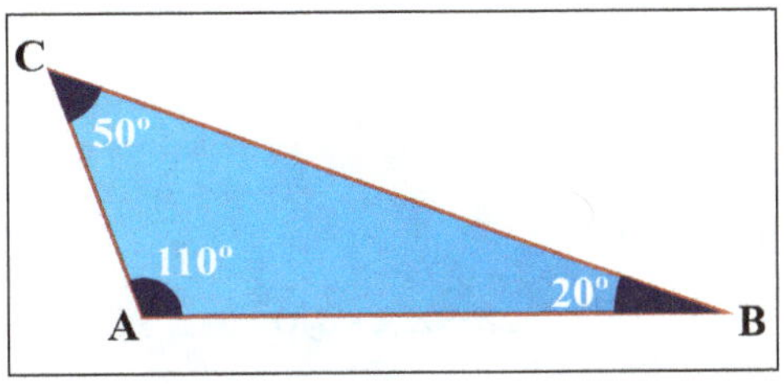

The side BC is the longest side which is opposite to the obtuse angle ∠A.

Example: A triangle cannot have more than one obtuse angle.

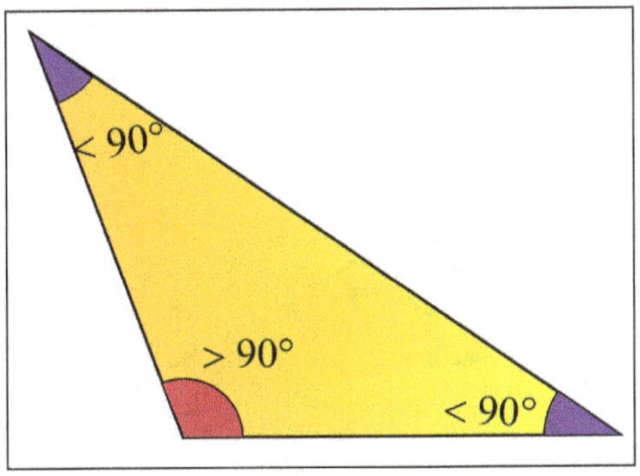

We know that the angles of any triangle add up to 180°. Consider the obtuse triangle shown above. We can observe that one of the angles measures greater than 90°, thus making it an obtuse angle. Even if we assume this obtuse angle to be 91°, the other two angles of the triangle will add up to 89° degrees. Hence, a triangle cannot have 2 obtuse angles.

Example: Sum of the other two angles in an obtuse-angled triangle is less than 90°. We just learnt that when one of the angles is an obtuse angle, the other two angles add up to less than 90°.

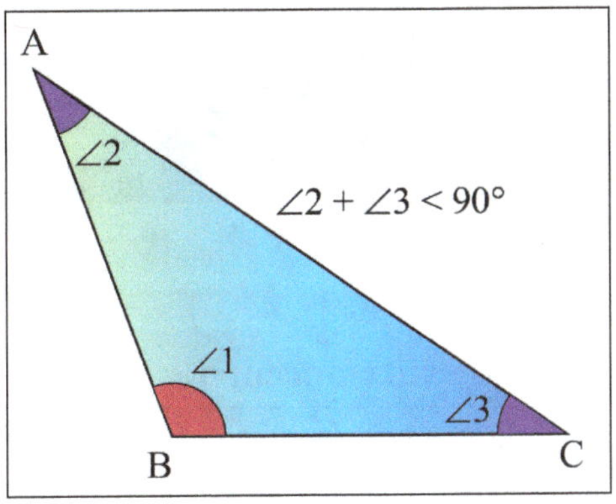

In the above triangle,

$$\angle 1 > 90°$$

We know that by angle sum property, the sum of the angles of a triangle is 180°.

Therefore,

$$\angle 1 + \angle 2 + \angle 3 = 180°$$

and,

$$\angle 1 > 90°$$

Subtracting the above two, we have:

$$\angle 2 + \angle 3 < 90°$$

Example: The circumcenter and the orthocenter of an obtuse-angled triangle lie outside the triangle. The orthocenter (H), the point at which all the altitudes of a triangle intersect, lies outside in an obtuse triangle.

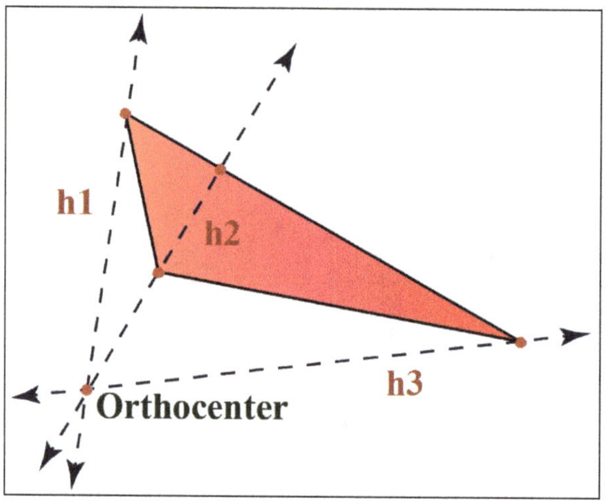

Circumcenter (O), the point which is equidistant from all the vertices of a triangle, lies outside in an obtuse triangle.

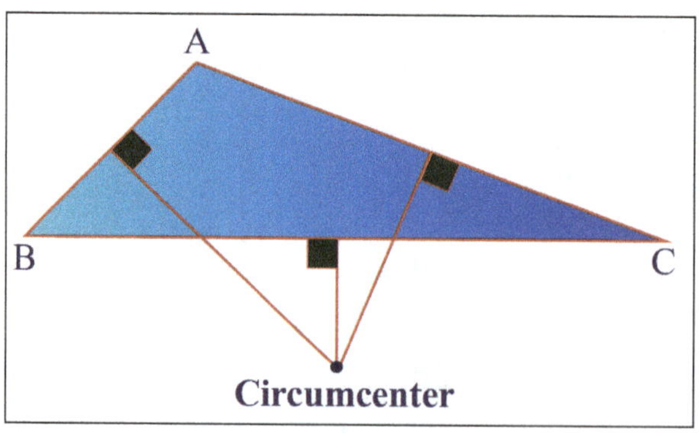

Solved Examples:

Example: Which of the following angle measures can form an obtuse triangle ABC?

a. 60°, 70°, 50°

b. 95°, 30°, 55°

c. 89°, 45°, 46°

d. 90°, 60°, 30°

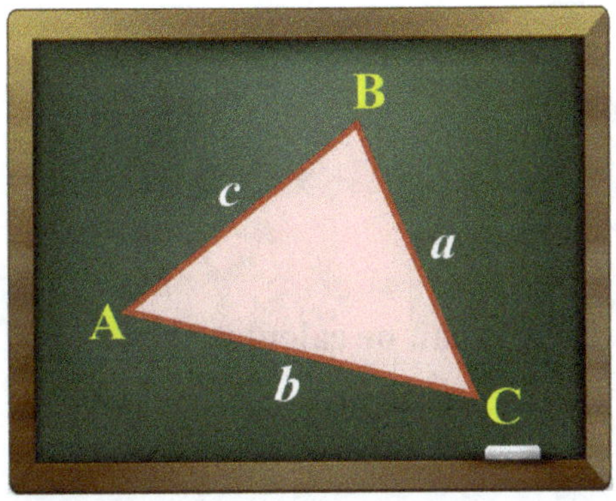

Solution: An obtuse triangle has one of the vertex angles as an obtuse angle (> 90°).

Among the given options, option (b) satisfies the condition.

∴ Option b forms an obtuse triangle.

Example: Find the area of an obtuse triangle whose base is 8 cm and height is 4 cm.

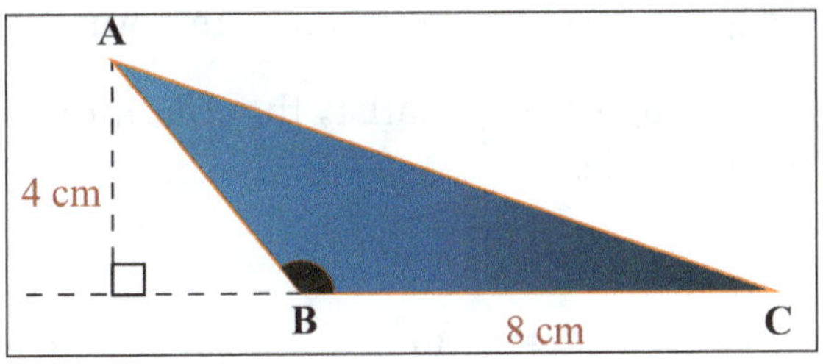

Solution:

Area of an obtuse-angled triangle,

$$\frac{1}{2} \times \text{base} \times \text{height}$$

Substituting the values of base and height, we get,

$$\text{Area} = \frac{1}{2} \times 8 \times 4 = 16 \text{cm}^2$$

∴ Area of the triangle = 16 cm².

Example: Find the height of the given obtuse triangle whose area = 60 cm² and base = 8 cm.

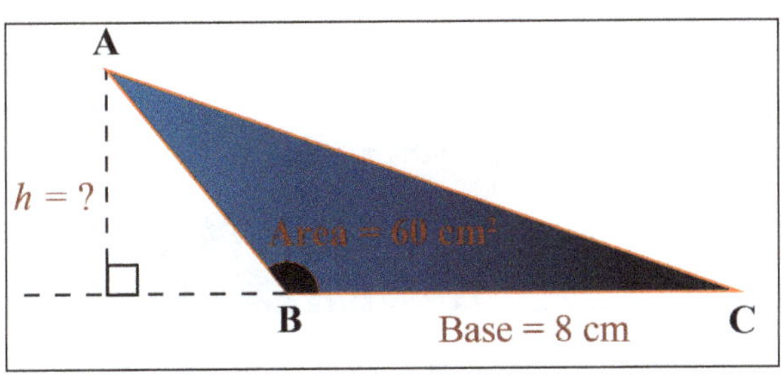

Solution:

Area of an obtuse-angled triangle,

$$\frac{1}{2} \times \text{base} \times \text{height}$$

Therefore, height of the obtuse triangle can be calculated by:

$$\text{Height} = \frac{2 \times \text{Area}}{\text{base}}$$

Substituting the values, we get:

$$\text{Height} = \frac{2 \times 60}{8} = 15\,\text{cm}$$

∴ Height = 15 cm.

Example: Can sides measuring 3 cm, 4 cm and 6 cm form an obtuse triangle?

Solution: The sides of an obtuse triangle should satisfy the condition that the sum of the squares of any 2 sides is greater than the third side.

We know that,

- a = 3 cm
- b = 4 cm
- c = 6 cm

Taking the squares of the sides, we get:

$$a^2 = 9, b^2 = 16 \text{ and } c^2 = 36$$

We know that,

$$36 > (9 + 16)$$

Therefore, the given measures can form the sides of an obtuse triangle. 3 cm, 4 cm and 6 cm can be the sides of an obtuse triangle.

Example: A triangle with one exterior angle measuring 80° is shown in the image. Can this triangle be an obtuse triangle?

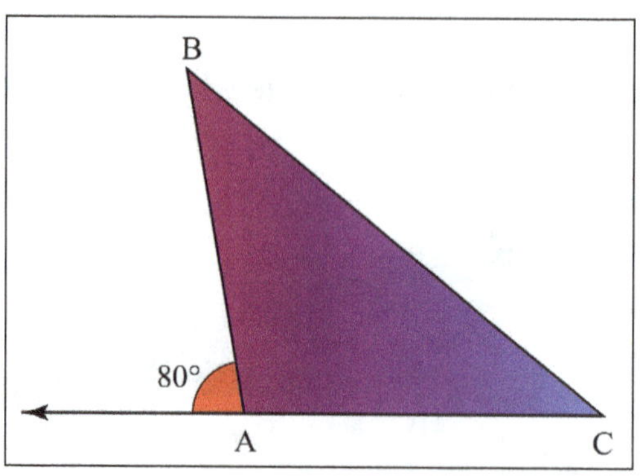

Solution: The exterior angle and the adjacent interior angle forms a linear pair (i.e , they add up to 180°).

Hence, the interior angle at vertex B is,

$$180 - 80 = 100^\circ$$

∴ The given triangle is an obtuse-angled triangle.

Centroid of Triangle

The centroid is the centre point of the object. The point in which the three medians of the triangle intersect is known as the centroid of a triangle. It is also defined as the point of intersection of all the three medians. The median is a line that joins the midpoint of a side and the opposite vertex of the triangle. The centroid of the triangle separates the median in the ratio of 2: 1. It can be found by taking the average of x-coordinate points and y-coordinate points of all the vertices of the triangle.

Centroid Theorem

The centroid theorem states that the centroid of the triangle is at 2/3 of the distance from the vertex to the mid-point of the sides.

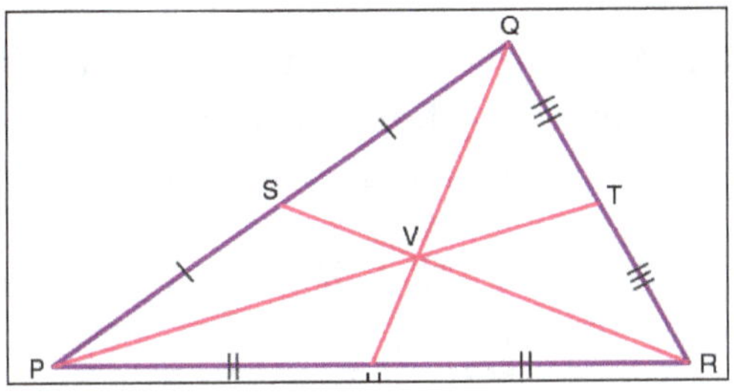

Suppose PQR is a triangle having a centroid V. S, T and U are the midpoints of the sides of the triangle PQ, QR and PR, respectively. Hence as per the theorem,

QV = 2/3 QU, PV = 2/3 PT and RV = 2/3 RS

Centroid of a Right Angle Triangle

The centroid of a right angle triangle is the point of intersection of three medians, drawn from the vertices of the triangle to the midpoint of the opposite sides.

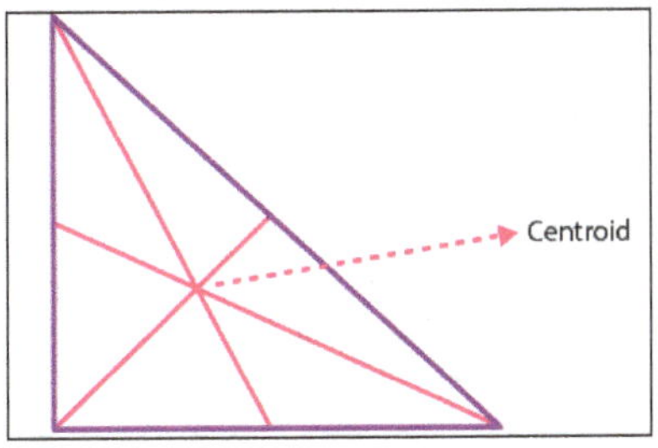

Centroid of a Square

The point where the diagonals of the square intersect each other is the centroid of the square. As we all know, the square has all its sides equal. Hence it is easy to locate the centroid in it. In the figure below, where O is the centroid of the square.

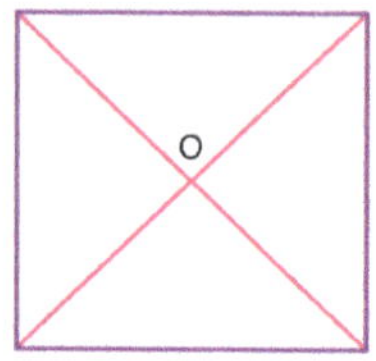

Properties of Centroid

The properties of the centroid are as follows:

- The centroid is the centre of the object.
- It is the centre of gravity.
- It should always lie inside the object.
- It is the point of concurrency of the medians.

Centroid Formula

Let's consider a triangle. If the three vertices of the triangle are $A(x_1, y_1)$, $B(x_2, y_2)$, $C(x_3, y_3)$, then the centroid of a triangle can be calculated by taking the average of X and Y coordinate points of all three vertices. Therefore, the centroid of a triangle can be written as:

Centroid of a triangle = $((x_1+x_2+x_3)/3, (y_1+y_2+y_3)/3)$

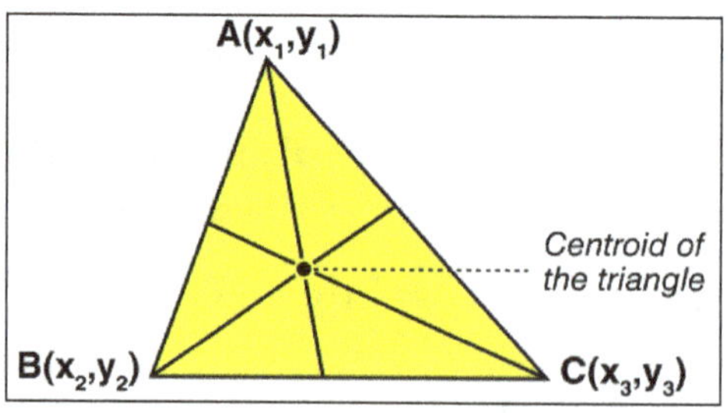

Centroid Formula for Different Shapes

Table: The list of centroid formula is given for different geometrical shapes.

Shapes	Figure	$\bar{x}$	$\bar{y}$	Area
Triangular area	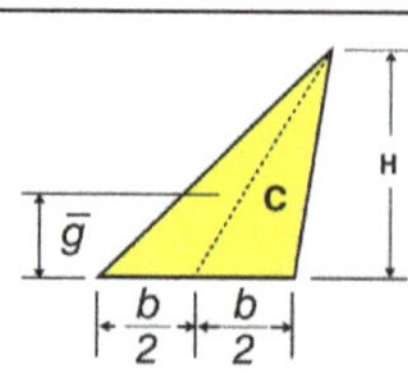	–	h/3	bh/2

Quarter-circular area	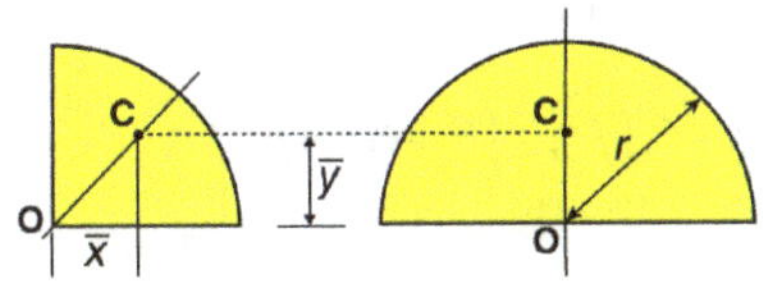	4r/3π	4r/3π	$\pi r^2/4$
Semicircular area	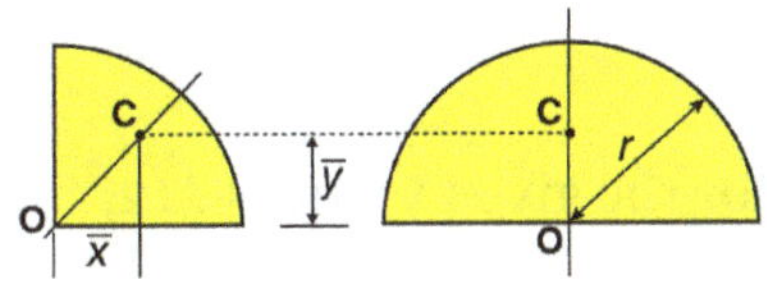	0	4r/3π	$\pi r^2/2$
Quarter-elliptical area	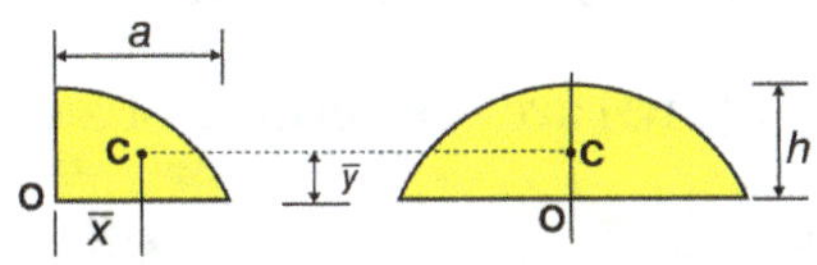	4a/3π	4b/3π	πab/4
Semi elliptical area	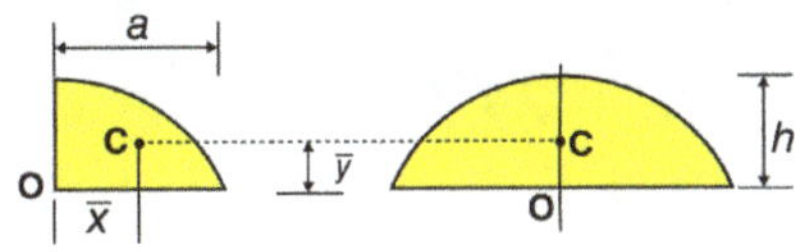	0	4b/3π	πab/2
Semiparabolic area	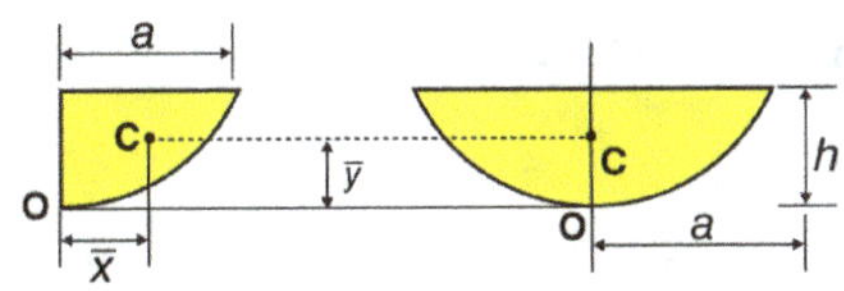	3a/8	3h/5	2ah/3
Parabolic area	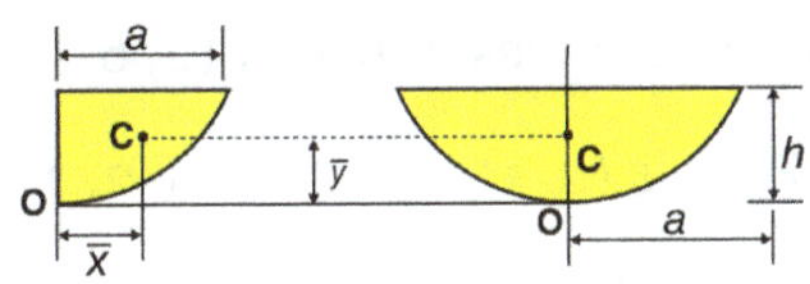	0	3h/5	4ah/3
Parabolic spandrel	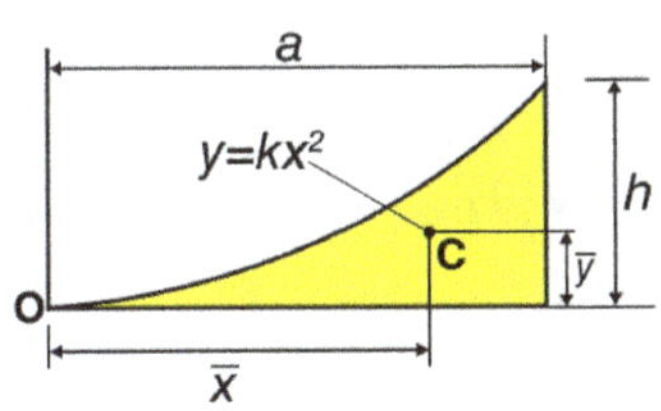	3a/4	3h/10	ah/3

Examples on Calculating Centroid

Find the solved examples below, to find the centroid of triangles with the given values of vertices.

Question: Find the centroid of the triangle whose vertices are A(2, 6), B(4, 9), and C(6,15).

Solution:

Given:

$$A(x_1, y_1) = A(2, 6)$$
$$B(x_2, y_2) = B(4,9)$$
$$C(x_3, y_3) = C(6,15)$$

We know that the formula to find the centroid of a triangle is $= ((x_1 + x_2 + x_3)/3, (y_1 + y_2 + y_3)/3)$.

Now, substitute the given values in the formula,

$$\text{Centroid of a triangle} = \left((2+4+6)/3,\ (6+9+15)/3\right)$$
$$= (12/3,\ 30/3)$$
$$= (4,\ 10)$$

Therefore, the centroid of the triangle for the given vertices A(2, 6), B(4,9), and C(6,15) is (4, 10).

Question: Find the centroid of the triangle whose vertices are A(1, 5), B(2, 6), and C(4, 10).

Solution: Given, A(1, 5), B(2, 6), and C(4, 10) are the vertices of a triangle ABC.

By the formula of the centroid we know,

$$\text{Centroid} = ((x_1 + x_2 + x_3)/3,\ (y_1 + y_2 + y_3)/3)$$

Putting the values, we get,

$$\text{Centroid} = (1+2+4)/3,\ (5+6+10)/3$$
$$=(\ 7/3,\ 21/3)$$
$$=(\ 7/3, 7)$$

Hence, the centroid of the triangle having vertices A(1, 5), B(2, 6), and C(4, 10) is (7/3, 7).

Question: If the vertices of a triangle PQR is (2, 1), (3, 2) and (-2, 4). Then find the centroid of it.

Solution: Given, (2, 1), (3, 2) and (-2, 4) are the vertices of triangle pQR.

By the formula of the centroid we know,

$$\text{Centroid} = ((x_1 + x_2 + x_3)/3,\ (y_1 + y_2 + y_3)/3)$$

Putting the values, we get;

$$\text{Centroid, O} = (2+3-2)/3,\ (1+2+4)/3$$
$$O = (3/3,\ 7/3)$$
$$O = (1,\ 7/3)$$

Hence, the centroid of the triangle having vertices (2, 1), (3, 2) and (-2, 4) is (1, 7/3).

References

- Triangles, introducing-geometry, pre-algebra, education: mathplanet.com, Retrieved 16, June 2020
- Equilateral-Triangle: mathworld.wolfram.com, Retrieved 12, April 2020

- Triangles-properties-types-geometry: mbacrystalball.com, Retrieved 09, March 2020
- Scalene-triangle, math-vocabulary, geometry: splashlearn.com, , Retrieved 13, August 2020
- Obtuse-triangles, geometry: cuemath.com, Retrieved 16, April 2020
- Acute-angle-triangle, Maths: byjus.com, Retrieved 14, May 2020

5

Basics of Circle

A circle is a closed, two-dimensional round shaped figure that has no corner or edges. While it may certainly be one of the easiest shapes to reproduce freehand, a better understanding of circles can be achieved only after mastering its basics. The different parts of a circle like the radius, chord or diameter, its perimeter, area and its properties are the topics addressed in this chapter.

A circle is all points in the same plane that lie at an equal distance from a center point. The circle is only composed of the points on the border. You could think of a circle as a hula hoop. It's only the points on the border that are the circle. The points within the hula hoop are not part of the circle and are called interior points.

The distance between the midpoint and the circle border is called the radius. A line segment that has the endpoints on the circle and passes through the midpoint is called the diameter. The diameter is twice the size of the radius. A line segment that has its endpoints on the circular border but does not pass through the midpoint is called a chord.

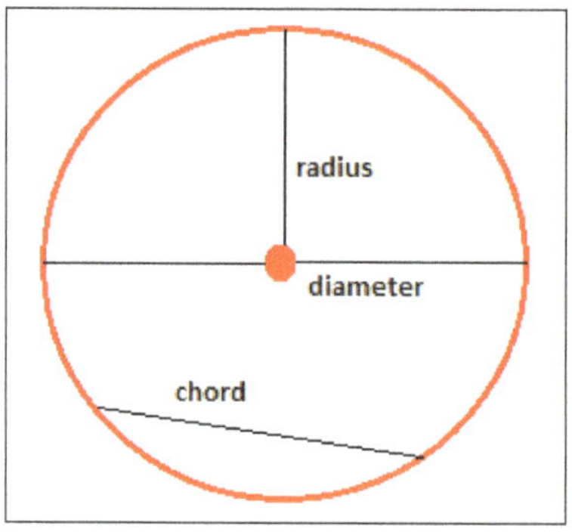

The distance around the circle is called the circumference, C, and could be determined either by using the radius, r, or the diameter, d:

$$/C = 2\pi r$$

$$C = \pi d$$

A circle is the same as 360°. You can divide a circle into smaller portions. A part of a circle is called an arc and an arc is named according to its angle. Arcs are divided into minor arcs (0° < v < 180°), major arcs (180° < v < 360°) and semicircles (v = 180°).

The length of an arc, l, is determined by plugging the degree measure of the Arc, v, and the circumference of the whole circle, C, into the following formula:

$$l = C \cdot \frac{v}{360}$$

When diameters intersect at the central of the circle they form central angles. Like when you cut a cake you begin your pieces in the middle.

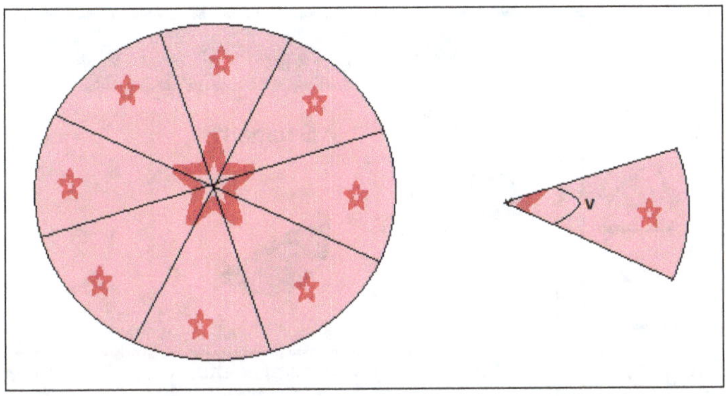

Example: As in the cake above we divide our circle into 8 pieces with the same angle. The circumference of the circle is 20 length units. Determine the length of the arc of each piece.

First we need to find the angle for each piece, since we know that a full circle is 360° we can easily tell that each piece has an angle of 360/8 = 45°. We plug these values into our formula for the length of arcs:

$$l = C \cdot \frac{v}{360}$$

$$l = 20 \cdot \frac{45}{360} = 2.5$$

Hence the length of our arcs is 2.5 length units. We could even easier have told this by simply dividing the circumference by the number of same size pieces: 20/8=2.5.

Parts of a Circle

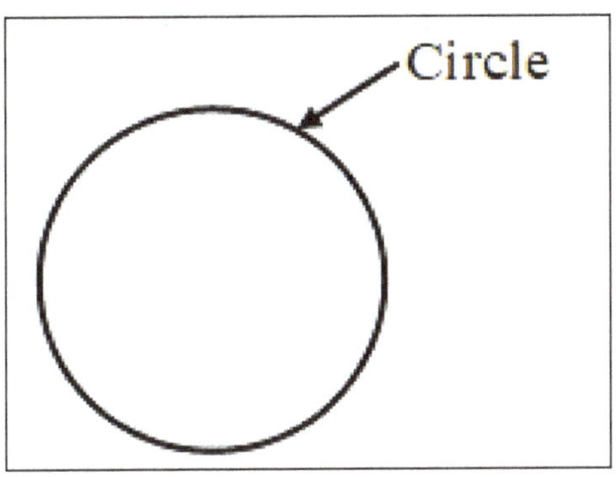

Circle is a closed figure that has a curvilinear boundary. When we think of circles, the very first thing that comes to our mind is its round shape, for example, bangles, coins, rings, plates, chapattis, pizzas, CDs etc. Wheels of a car, bus, cycle, truck, train, and aeroplanes are also round in shape.

If we take a stone, tie it to one end of a string and swing it in the air by holding the other end of the string, the path traced by the stone will be a circular path and it will make a circle.

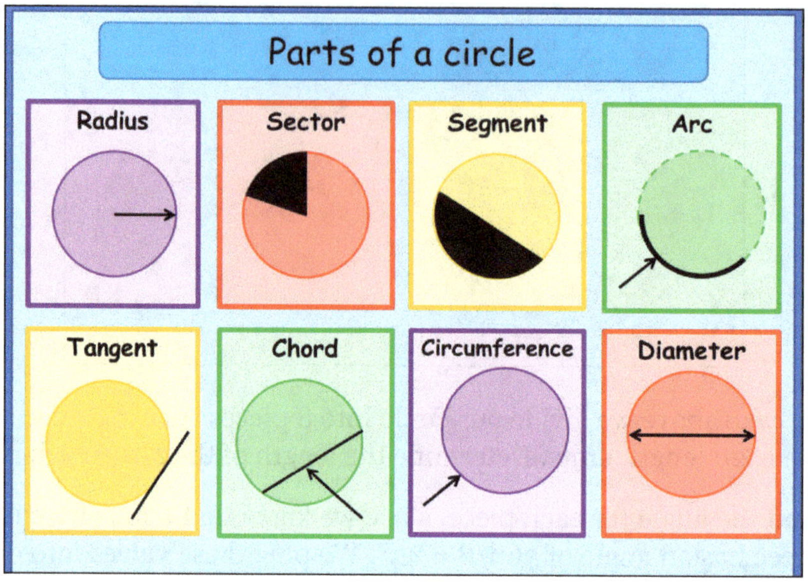

Circle: A circle is a collection of all those points in a plane that are at a given constant distance from a given fixed point in the plane.

Centre: Circle is a closed figure made up of points in a plane that are at the same distance from a fixed point, called the centre of the circle. In the figure O is the centre.

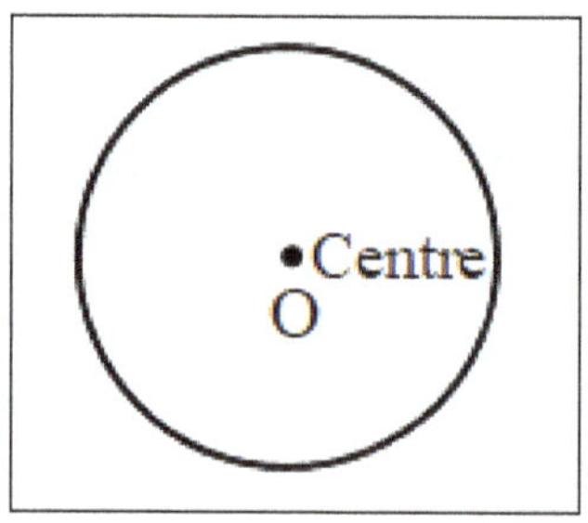

Radius: The constant distance from its centre is called the radius of the circle. In the figure, OA is radius.

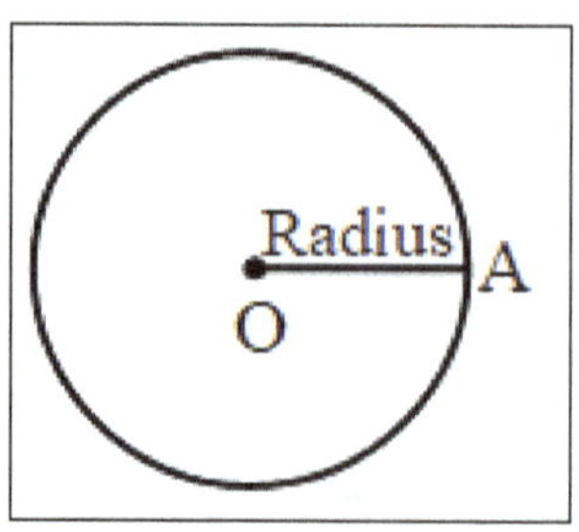

Chord: A line segment joining two points on a circle is called a chord of the circle. In the figure, AB is a chord of the circle. If a chord passes through centre then it is longest chord. PQ, PR, and ST are chords of the circle. Chord ST passes through the centre, hence it is a diameter.

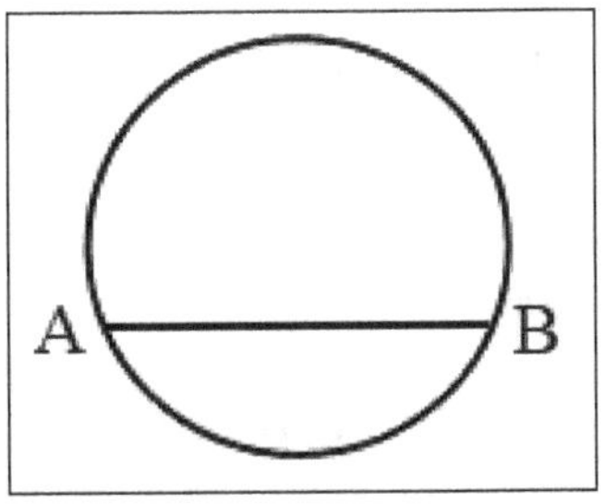

Diameter: A chord passing through the centre of a circle is called the diameter of the circle. A circle has an infinite number of diameters. CD is the diameter of the circle as shown in the figure. If d is the diameter of the circle then d = 2r. Where r is the radius or the longest chord is called diameter. In the figure, AB is the diameter and the arcs CD and DC are semicircles.

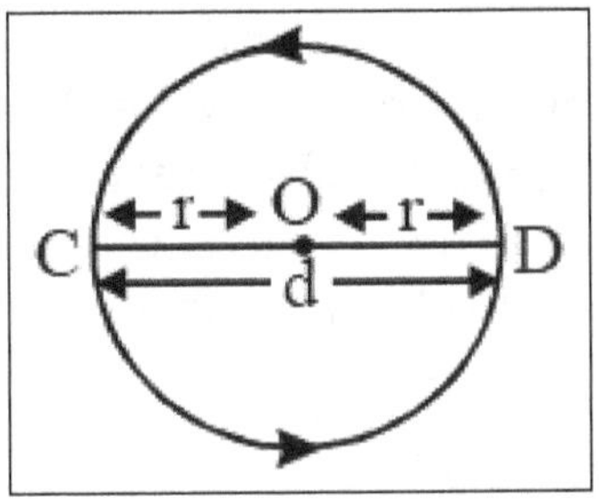

Arc: A continuous piece of a circle is called an arc. Let A, B, C, D, E, F be the points on the circle. The circle is divided into different pieces. Then, the pieces AB, BC, CD, DE, EF etc. are all arcs of the circle.

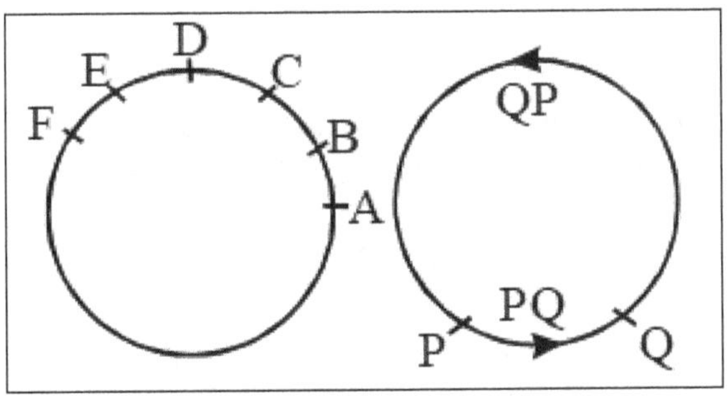

Let P, Q be two points on the circle. These P, Q divides the circle into two parts. Each part is an arc. These arcs are denoted in anti-clockwise direction,

From P to Q as $\overset{\frown}{PQ}$ and form Q to P as $\overset{\frown}{QP}$.

The counter clockwise direction distinguishes between these two arcs $\overset{\frown}{PQ}$ and $\overset{\frown}{QP}$.

The length of arc $\overset{\frown}{PQ}$ can be less than, equal to or greater than the length of the arc $\overset{\frown}{QP}$ i.e.,

- $l\left(\overset{\frown}{PQ}\right) < l\left(\overset{\frown}{QP}\right)$
- $l\left(\overset{\frown}{PQ}\right) = l\left(\overset{\frown}{QP}\right)$
- $l\left(\overset{\frown}{PQ}\right) > l\left(\overset{\frown}{QP}\right)$

When $l(\overparen{PQ}) < l(\overparen{QP})$ then the arc $(\overparen{PQ})$ is called a minor arc.

If $l(\overparen{PQ}) = l(\overparen{QP})$ then the arc $\overparen{PQ}$ and $\overparen{QP}$ are called semi circle.

At this time points of arc at end of diameter.

And when $l(\overparen{PQ}) > l(\overparen{QP})$ then the arc $\overparen{PQ}$ is called a major arc.

Circumference of a circle: The perimeter of a circle is called its circumference. The circumference of a circle of radius r is $2\pi r$.

Semicircle: The diameter of a circle divides the circle into two equal parts. Each part is called a semi-circle. We can also say that half of a circle is called a semi-circle. In the figure, AXB and AYB represent two semi-circles.

Segment: Let AB be a chord of the circle. Then, AB divides the region enclosed by the circle (i.e., the circular disc) into two parts. Each of the parts is called a segment of the circle. The segment, containing the minor arc is called minor segment and the segment, containing the major arc, is called the major segment and segment of a circle is the region between an arc and chord of the circle.

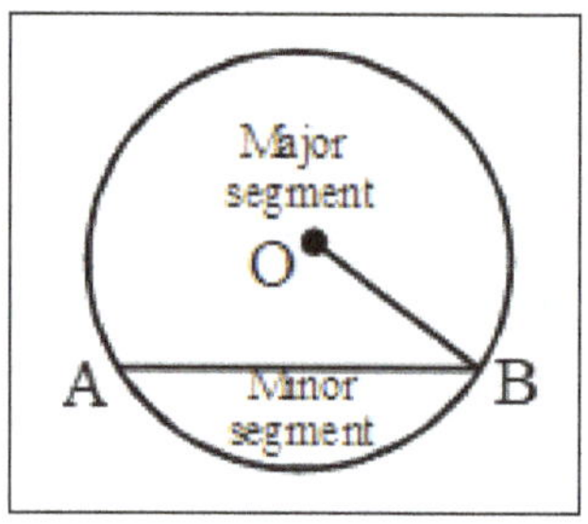

Central Angles: Consider a circle - The angle subtended by an arc at the centre O is called the central angle. The vertex of the central angle is always at the centre O.

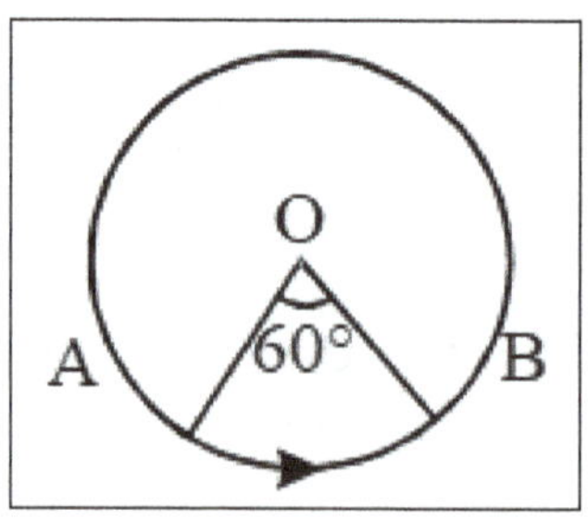

Degree measure of an arc: Degree measure of a minor arc is the measure of the central angle subtended by the arc.

In the figure, the measure of the arc $\overparen{PQ}$ is 60° i.e. $m\overparen{PQ} = 60°$.

The measure of a major arc is $360° - m\overparen{PQ}$ the degree measure of the corresponding minor arc.

The degree measure of the major arc is $360° - 60° = 300°$.

$\therefore \quad m\overparen{QP} = 300°$.

The degree measure of the circumference of the circle is always 360°.

Interior and Exterior of Circle: A circle divides the plane on which lies into three parts:

- Inside the circle which is called the interior of the circle.
- Circle.
- Outside the circle, which is called the exterior of the circle.

The circle and its interior make up the circular region.

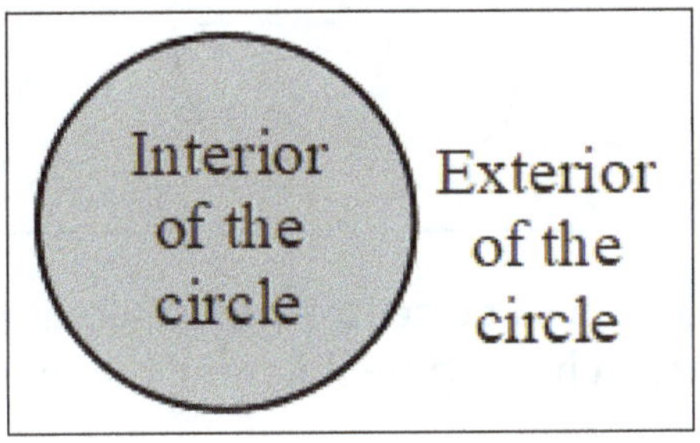

Sector: A sector is that region of a circular disc which lies between an arc and the two radii joining the extremities of the arc and the centre. OAB is a sector as shown in the figure.

Quadrant: One fourth of a circular disc is called a quadrant.

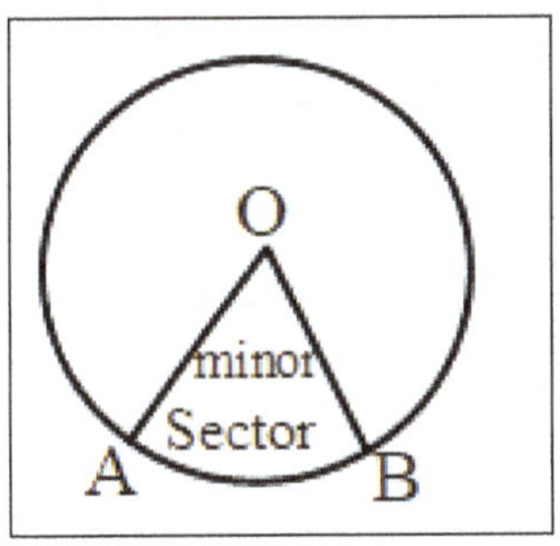

Position of a point:

- Point inside the circle: A point P, such that OP < r, is said to lie inside the circle.

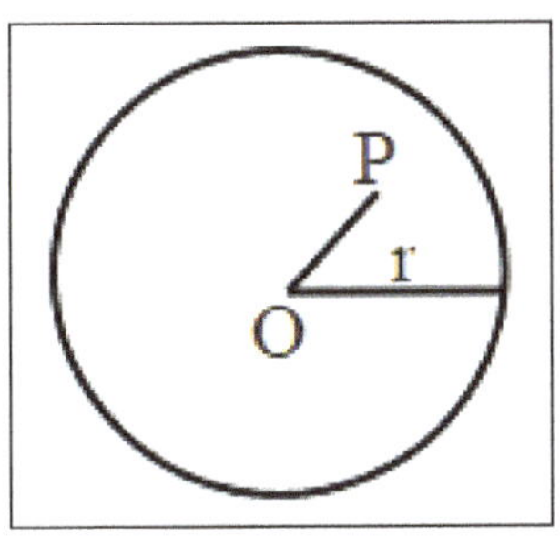

- The point inside the circle is also called interior point. (Example: Centre of circle).
    - Point outside the circle: A point Q, such that OQ > r, is said to lie outside the circle C (O, r) = {X, OX = r}. The point outside the circle is also called exterior point.

- Point on the circle: A point S, such that OS = r is said to lie on the circle C(O, r) = {X, OX = r}.
- Circular disc: It is defined as a set of interior points and points on the circle. In set notation, it is written as: C(O, r) = {X : P OX ≤ r}.

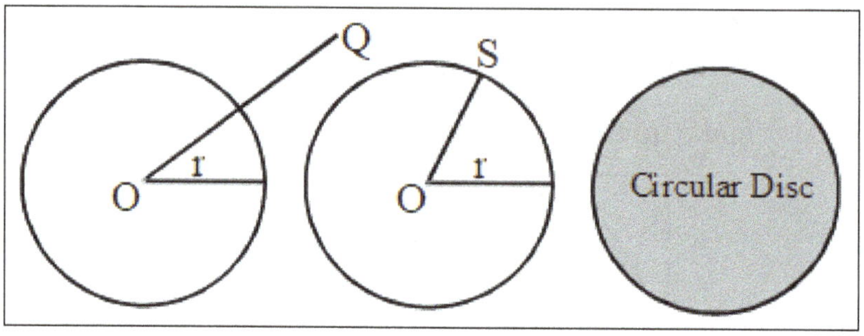

Concentric Circles: Circles having the same centre and different radius are said to be concentric circles. The word 'radius' is used for a line segment joining the centre to any point on the circle and also for its length.

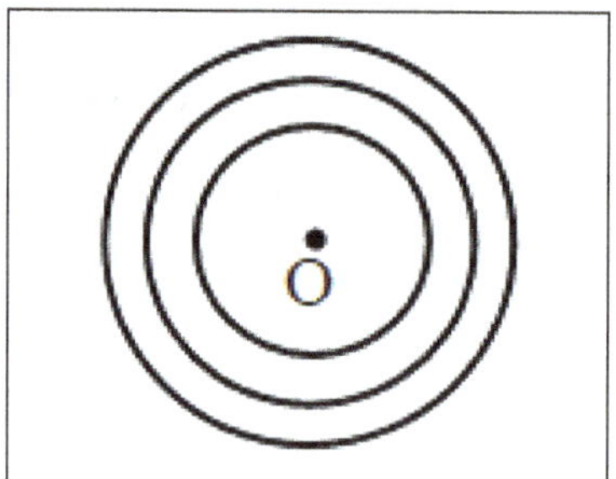

Concentric Circles

Congruence of Circles and Arcs:

- Congruent circles: Two circles are said to be congruent if and only if, one of them can be superposed on the other, so as the cover it exactly. It means two circles are congruent if and only if, their radii are equal. i.e., C (O, r) and C (O', r) are congruent if only if r = s.

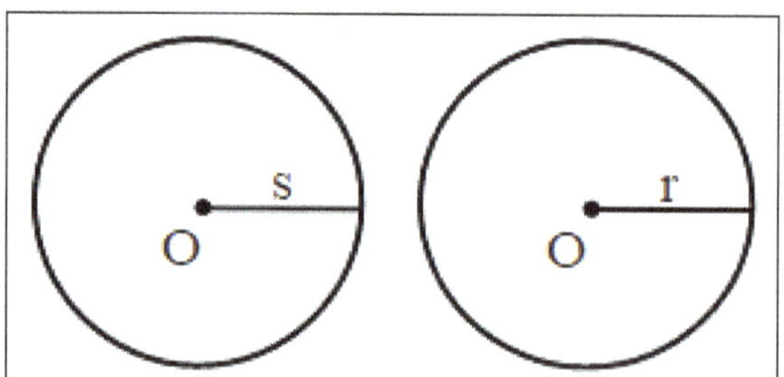

Congruent circles if r = s

- Congruent arcs: Two arcs of a circle are congruent, if either of them can be superposed on the other, so as to cover it exactly. It is only possible, if degree measure of two arcs is the same.

Example: Take two points A and B on a plane sheet. Draw a circle with A as a centre, AC as radius and B in its exterior.

Solution: Mark two points A and B on a paper.

- A
- B

As the point B should be in the exterior of the circle, take A as the centre and radius (r) less than AB to draw a circle.

Example: Find the diameter of the circle of radius 6 cm.

Solution: We know,

Diameter = 2 × radius

∴ Diameter = 2 × 6 cm = 12 cm

Perimeter of a Circle

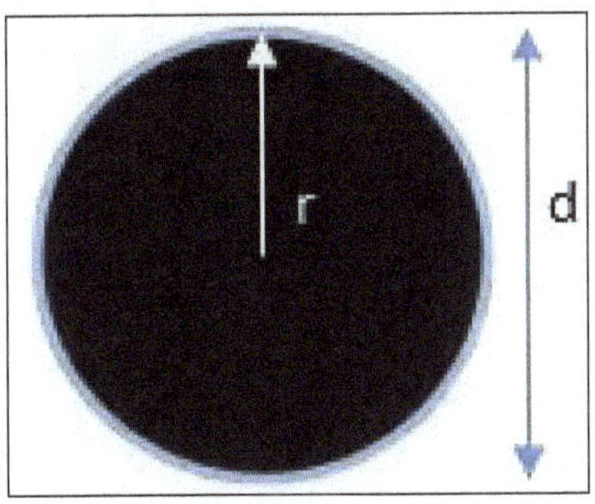

Perimeter is the distance around a closed figure and is typically measured in millimetres (mm), centimetres (cm), metres (m) and kilometres (km). These units are related as follows:

10 mm = 1 cm

100 cm = 1 m

1000 m = 1 km

The word 'perimeter' is also sometimes used instead of circumference.

If we know the Radius

Given the radius of a circle, the circumference or perimeter can be calculated using the formula below:

$$\text{Perimeter (P)} = 2 \cdot \pi \cdot R$$

where:

R is the radius of the circle

π is Pi, approximately 3.142

If we know the Diameter

If we know the diameter of a circle, the circumference can be found using the formula,

$$\text{Perimeter (P)} = \pi \cdot D$$

where:

D is the diameter of the circle

π is Pi, approximately 3.142

If we know the Area

If we know the area of a circle, the circumference can be found using the formula:

$$\text{Perimeter (P)} = \sqrt{(4 \cdot \pi \cdot A)}$$

where:

A is the area of the circle

π is Pi, approximately 3.142

Example: A circular flower-bed has a radius of 9 m. Find the perimeter/circumference of the flower-bed.

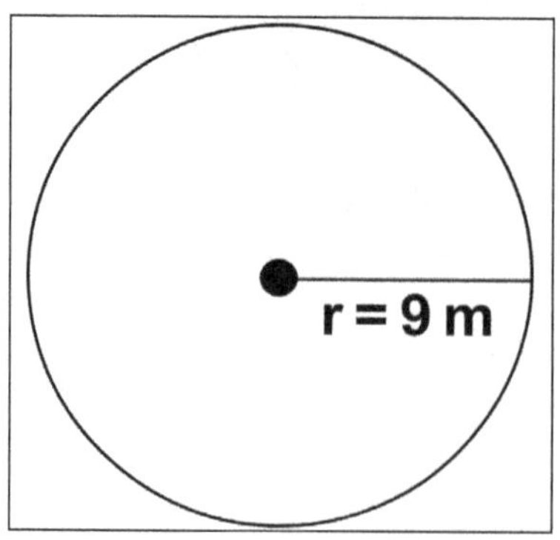

Solution:

$$P = 2 \cdot \pi \cdot R$$

$$P = 2 \cdot 3.1416 \cdot 9$$

$$P = 56.5487 \text{ cm}$$

So, the perimeter/circumference of the flower-bed is 56.5487 m.

Example: Find the perimeter of the given circle whose diameter is 4.4 cm.

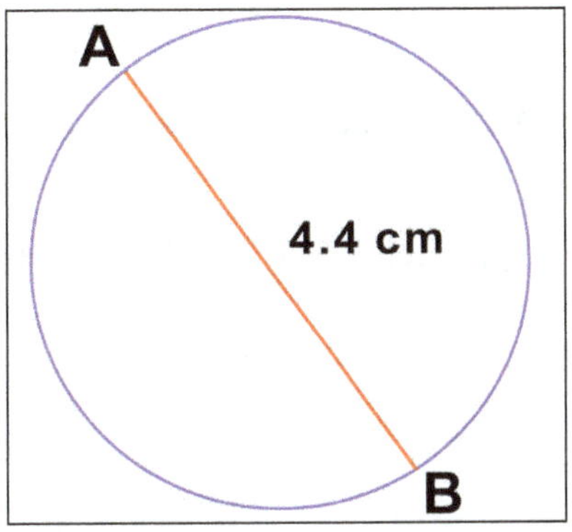

Solution:

Given that: Diameter of the circle (D) = 4.4 cm.

We know the formula to find the perimeter of the circle if the diameter is given, namely $\pi \cdot D$.

Substitute the diameter 4.4 and Pi value as 3.14 in the above formula.

Perimeter = (3.14)(4.4) = 13.82

Therefore 13.82 cm is the perimeter of the given circle.

Example: If the radius is 11.7 cm. Find perimeters (circumference) of the circle.

Solution:

Given that: Radius (r) = 11.7cm

Perimeter (circumference) of circle $P = 2 \pi r$

Substitute the r value in the formula, we have:

P = 2 x 3.14 x 11.7

P = 79.56 cm

Thus, the perimeter of the circle is 79.56cm.

Example: Find the perimeter and area of the circle, if the radius of the circle is 8cm.

Solution: We have given the radius, which is 8cm. So, by using the formula of the perimeter of the circle, we have:

$P = 2\pi r$

$P = 2 \times 3.14 \times 8$

P = 50.24 cm

And for the area of the circle:

$A = \pi r^2$

$$A = 3.14 \times (8)^2$$

$$A = 200.96 cm^2$$

Example: The wheel of a bullock cart has a radius of 6 m. If the wheel rotates once how much distance does the cart move?

Solution: If the wheel rotates once, the cart will move by a distance equal to the perimeter of the wheel.

Step:

$$P = 2\pi r$$

$$P = 2 \times 3.14 \times 6 = 37.68 \text{ m}$$

Thus, the bullock cart moves 37.68 m in one revolution of the wheel.

Area of a Circle

Area of a circle is the region occupied by the circle in a two-dimensional plane. It can be determined easily using a formula, $A = \pi r^2$, (Pi r-squared) where r is the radius of the circle. The unit of area is the square unit, such as m^2, cm^2, etc.

$$\text{Area of Circle} = \pi r^2 \text{ or } \pi d^2/4, \text{ (square unit) where } \pi = 22/7 \text{ or } 3.14$$

This area formula is useful for measuring the space occupied by a circular field or a plot. Suppose, if you have the plot to fence it, then the area formula will help you to check how much fencing is required. Or suppose if you have to buy a table cloth, then how much portion of cloth is needed to cover it completely.

A circle closed plane geometric shape. In technical terms, a circle is a locus of a point moving around a fixed point at a fixed distance away from the point. Basically, a circle is a closed curve with its outer line equidistant from center. The fixed distance from the point is the radius of the circle. In real life, you will get many examples of the circle such as a wheel, pizzas, a circular ground, etc. Now let us learn, what are the terms used in the case of a circle.

Radius

The radius of the circle is the line which joins the centre of the circle to the outer boundary. It is usually represented by 'r' or 'R'. In the formula for area and circumference of a circle, radius plays an important role.

Diameter

The diameter of the circle is the line which divides the circle into two equal parts. In an easy way we can say, it is just the double of the radius of the circle and is represented by 'd' or 'D'. Therefore,

$$d = 2r \text{ or } D = 2R$$

If the diameter of the circle is known to us, we can calculate the radius of the circle, such as,

$$r = d/2 \text{ or } R = D/2$$

Area of Circle Definition

Any geometrical shape has its own area. This area is the region occupied the shape in a two-dimensional plane. Now we will learn about the area of the circle. So the area covered by one complete cycle of the radius of the circle on a two-dimensional plane is the area of that circle. Now how can we calculate the area for any circular object or space? In this case, we use the formula for the circle's area.

Area of a Circle Formula

Let us take a circle with radius r.

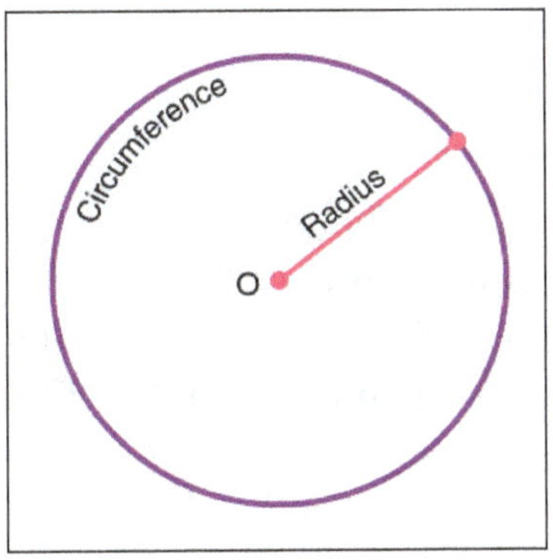

In the above figure, we can see a circle, where radius r from the centre 'o' to the boundary of the circle. Then the area for this circle, A, is equal to the product of pi and square of the radius. It is given by,

$$\text{Area of a Circle, } A = \pi r^2 \text{ square units}$$

Here, the value of pi, $\pi = 22/7$ or 3.14 and r is the radius.

Derivation of Area of Circle

Area of a circle can be visualized and proved using two methods, namely:

- Determining the circle's area using rectangles.
- Determining the circle's area using triangles.

Using Areas of Rectangles

The circle is divided into 16 equal sectors, and the sectors are arranged. The area of the circle will be equal to that of the parallelogram-shaped figure formed by the sectors cut out from the circle. Since the sectors have equal area, each sector will have equal arc length. The red coloured sectors will contribute to half of the circumference, and blue coloured sectors will contribute to the other half. If the number of sectors cut from the circle is increased, the parallelogram will eventually look like a rectangle with length equal to πr and breadth equal to r.

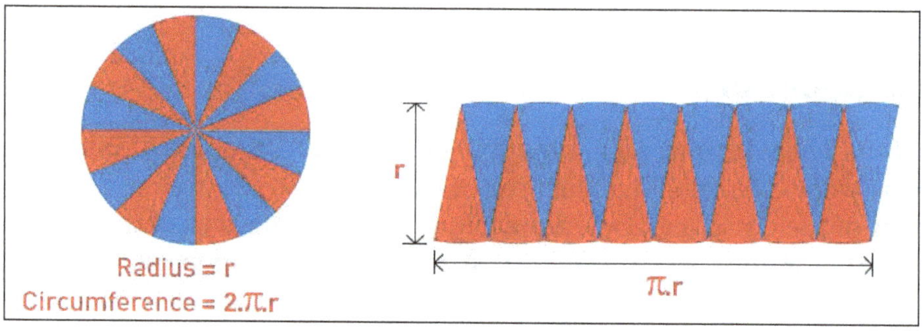

Visualizing area of circle using area of rectangle.

The area of a rectangle (A) will also be the area of a circle. So, we have,

$$A = \pi r \times r$$

$$A = \pi r^2$$

Using Areas of Triangles

Fill the circle with radius r with concentric circles. After cutting the circle along the indicated line in figure below and spreading the lines, the result will be a triangle. The base of the triangle will be equal to the circumference of the circle, and its height will be equal to the radius of the circle.

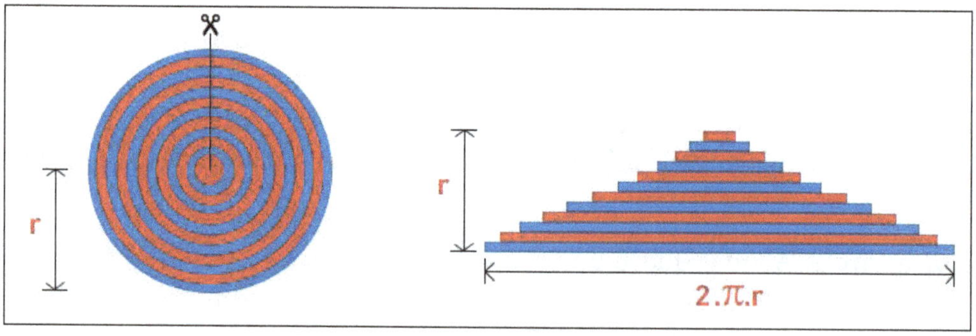

Visualizing area of circle using area of triangle.

So, the area of the triangle (A) will be equal to the area of the circle. We have,

$$A = \frac{1}{2} \times \text{base} \times \text{height}$$

$$A = 1/2 \times (2\pi r) \times r$$

$$A = \pi r^2$$

Surface Area of Circle Formula

A circle is nothing but the 2-D representation of a sphere. The total area that is taken inside the boundary of the circle is the surface area of the circle. When we say we want the area of the circle, then we mean the surface area of the circle itself. Sometimes, the volume of a circle also defines the area of a circle.

When the length of the radius or diameter or even the circumference of the circle is already given, then we can use the surface formula to find out the surface area. The surface is represented in square units.

The surface area of the circle = A = $\pi \times r^2$.

Circumference of Circle

A perimeter of closed figures is defined as the length of its boundary. When it comes to circles, the perimeter is given in a different name. It is called the "Circumference" of the circle. This circumference is the length of the boundary of the circle. If we open the circle to form a straight line, then the length of the straight line is the circumference. To define the circumference of the circle, knowledge of a term known as 'pi' is required. Consider the circle shown in the figure below, with centre at O and radius r.

A circle and a rope.

The perimeter of the circle is equal to the length of its boundary. The length of rope which wraps around its boundary perfectly will be equal to its circumference, which can be measured by using the formula:

Circumference / Perimeter = $2\pi r$ units

where r is the radius of the circle.

π, read as 'pi' is defined as the ratio of the circumference of a circle to its diameter. This ratio is the same for every circle. Consider a circle with radius 'r' and circumference 'C'. For this circle,

- π = Circumference/Diameter
- $\pi = C/2r$
- $C = 2\pi r$

The same is shown in figure below:

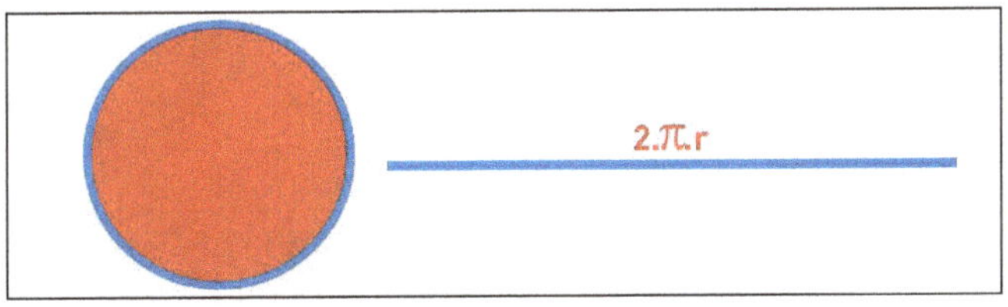

Depiction of length of boundary of circle.

Calculation of Area of Circle

As we know, the area of circle is equal to pi times square of its radius, i.e. $\pi \times r^2$. To find the area of circle we have to know the radius or diameter of the circle. For example, if the radius of circle is 7cm, then its area will be:

Area of circle with 7 cm radius = $\pi r^2 = \pi(7)^2$

$= 22/7 \times 7 \times 7 = 22 \times 7 = 154$ sq.cm.

Also, if we know the circumference of the circle, then we can find the area of circle. How? Since, the circumference is 2 times of product of pi and radius of circle, such as:

$C = 2\pi r$

Therefore, here we can find the value of radius,

$r = C/2\pi$

Once, we have evaluated the value of radius, we can easily find the area.

Square Area and Circle Area

The area of circle is estimated to be the 80% of area of square, when the diameter of circle and length of side of square is same. Student can also do an activity by inserting a circular object into a square shape with same diameter and side-length, respectively. If area of square is 100 sq.unit, then the area of circle will be approximately 80 sq.unit of it.

Properties of a Circle

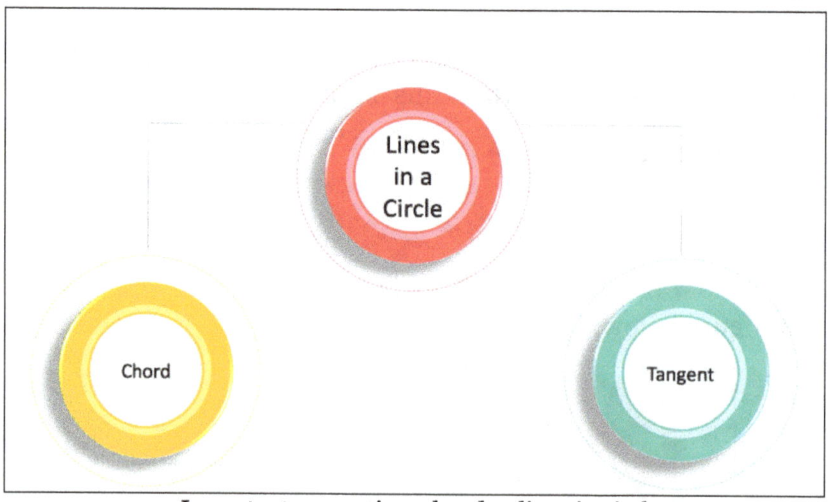

Important properties related to lines in circle.

Properties Related to Lines in a Circle

Chord

A chord is a line segment whose endpoints lie on the boundary of the circle.

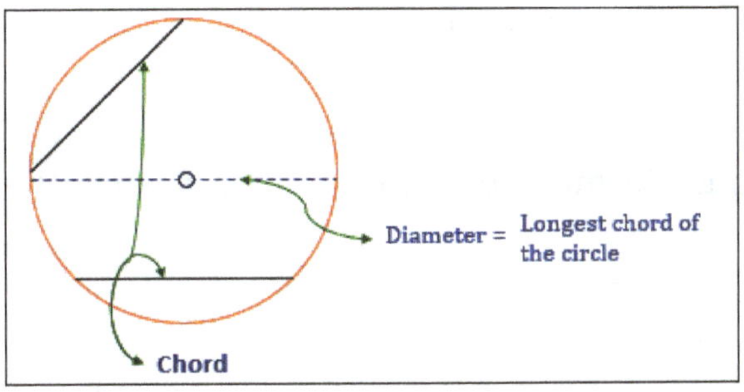

Properties of Chord

Perpendicular dropped from the center divides a chord into two equal parts.

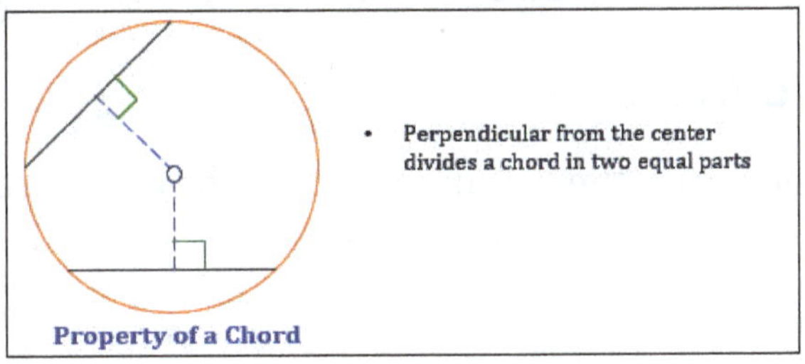

Tangent

Tangent is a line that touches the circle at any point.

Properties of Tangent

Radius is always perpendicular to the tangent at the point where it touches the circle.

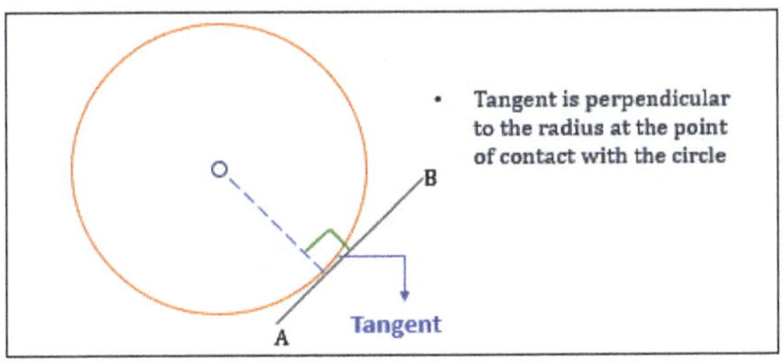

Important Properties of Circle – Related to Angles

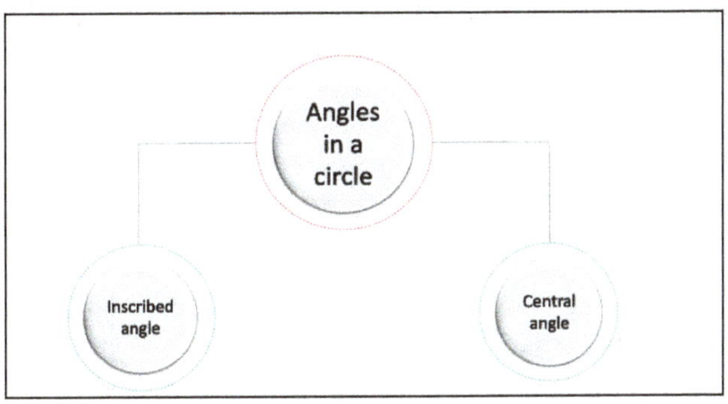

Important properties related to angles in a circle.

Properties Related to Angles in a circle

Inscribed Angle

An inscribed angle is the angle formed between two chords when they meet on the boundary of the circle.

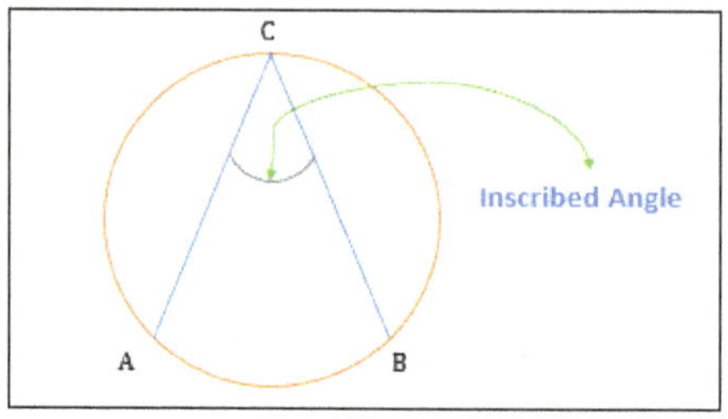

Properties of Inscribed Angles

1) Angles formed by the same arc on the circumference of the circle are always equal.

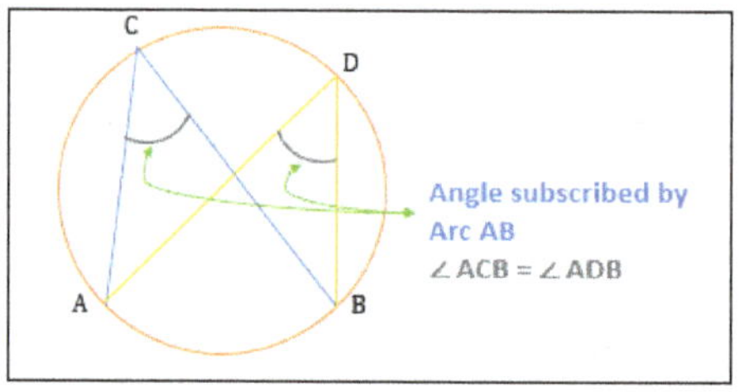

2) The angle in a semi-circle is always 90°.

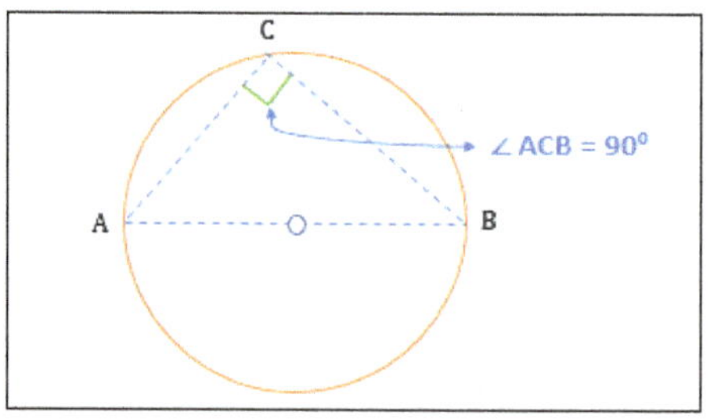

Central Angle

A central angle is the angle formed when two-line segments meet such that one of the endpoints of both the line segment is at the center and another is at the boundary of the circle.

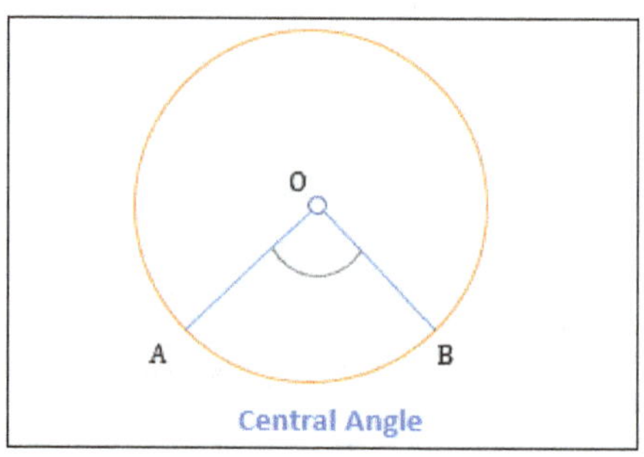

Property of Central Angles

An angle formed by an arc at the center is twice the inscribed angle formed by the same arc.

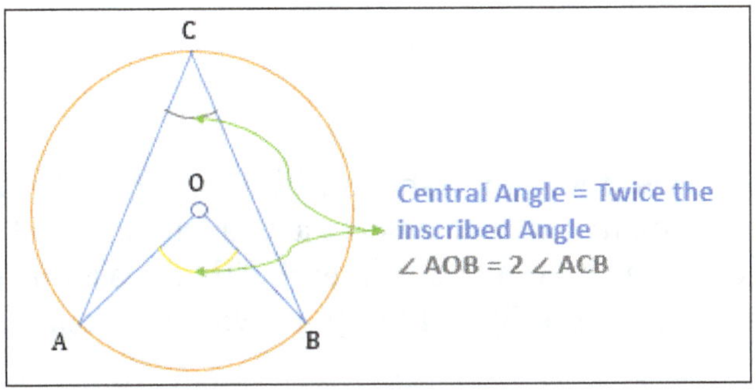

Important Circle Formulas: Area and Perimeter

The following are some mathematical formulae that will help you calculate the area and perimeter/circumference of a circle.

Perimeter:

- Perimeter or the Circumference of the circle = $2 \times \pi \times R$.
- Length of an Arc = (Central angle made by the arc/360°) $\times 2 \times \pi \times R$.

Area:

- Area of the circle = $\pi \times R^2$.
- Area of the sector = (Central angle made by the sector/360°) $\times \pi \times R^2$.

References

- Basic-information-about-circles, education, geometry: mathplanet.com, Retrieved 14, May 2020
- Parts-of-the-circle: aplustopper.com, Retrieved 19, March 2020
- Geometry-Perimeter-of-Circle, Math-Formulas: web-formulas.com, Retrieved 17, July 2020
- Area-circle, maths: byjus.com, Retrieved 10, August 2020
- Circle-properties-circle-formulas-circle-area-circle-circumference: e-gmat.com, Retrieved 14, February 2020

6

Types of Quadrilaterals

A quadrilateral is a closed, two-dimensional shape with four straight sides. A quadrilateral with four equal sides is called a square. A rectangle is a quadrilateral with four right angles and parallel opposite sides of equal lengths. A quadrilateral with at least one pair of parallel sides is a trapezoid. This chapter has been carefully written to provide a comprehensive understanding of the subject matter.

Quadrilateral

In Euclidean plane geometry, a quadrilateral is a polygon with four edges (or sides) and four vertices or corners. Sometimes, the term quadrangle is used, by analogy with triangle, and sometimes tetragon for consistency with pentagon (5-sided), hexagon (6-sided) and so on.

Some types of quadrilaterals.

The origin of the word "quadrilateral" is the two Latin words *quadri*, a variant of four, and *latus*, meaning "side".

Quadrilaterals are simple (not self-intersecting) or complex (self-intersecting), also called crossed. Simple quadrilaterals are either convex or concave.

The interior angles of a simple (and planar) quadrilateral *ABCD* add up to 360 degrees of arc, that is,

$$\angle A + \angle B + \angle C + \angle D = 360^{\circ}.$$

This is a special case of the n-gon interior angle sum formula $(n - 2) \times 180°$.

All non-self-crossing quadrilaterals tile the plane by repeated rotation around the midpoints of their edges.

Simple Quadrilaterals

Any quadrilateral that is not self-intersecting is a simple quadrilateral.

Convex Quadrilaterals

In a convex quadrilateral, all interior angles are less than 180° and the two diagonals both lie inside the quadrilateral.

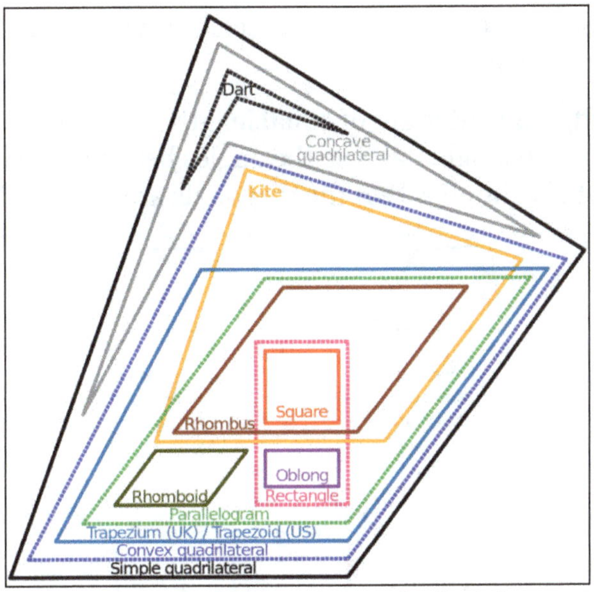

Euler diagram of some types of simple quadrilaterals. (UK) denotes British English and (US) denotes American English.

- Irregular quadrilateral (British English) or trapezium (North American English): no sides are parallel. (In British English this was once called a *trapezoid*).
- Trapezium (UK) or trapezoid (US): at least one pair of opposite sides are parallel.
- Isosceles trapezium (UK) or isosceles trapezoid (US): one pair of opposite sides are parallel and the base angles are equal in measure. Alternative definitions are a quadrilateral with an axis of symmetry bisecting one pair of opposite sides, or a trapezoid with diagonals of equal length.
- Parallelogram: a quadrilateral with two pairs of parallel sides. Equivalent conditions are that opposite sides are of equal length; that opposite angles are equal; or that the diagonals bisect each other. Parallelograms also include the square, rectangle, rhombus and rhomboid.
- Rhombus or rhomb: all four sides are of equal length. An equivalent condition is that the diagonals perpendicularly bisect each other. Informally: "a pushed-over square" (but strictly including a square too).
- Rhomboid: a parallelogram in which adjacent sides are of unequal lengths and angles are oblique (not right angles). A parallelogram which is not a rhombus. Informally: "a pushed-over oblong" (but strictly including an oblong too).

- Rectangle: all four angles are right angles. An equivalent condition is that the diagonals bisect each other and are equal in length. Informally: "a box or oblong" (including a square).
- Square (regular quadrilateral): all four sides are of equal length (equilateral), and all four angles are right angles. An equivalent condition is that opposite sides are parallel (a square is a parallelogram), that the diagonals perpendicularly bisect each other, and are of equal length. A quadrilateral is a square if and only if it is both a rhombus and a rectangle (four equal sides and four equal angles).
- Oblong: a term sometimes used to denote a rectangle which has unequal adjacent sides (i.e. a rectangle that is not a square).
- Kite: two pairs of adjacent sides are of equal length. This implies that one diagonal divides the kite into congruent triangles, and so the angles between the two pairs of equal sides are equal in measure. It also implies that the diagonals are perpendicular.

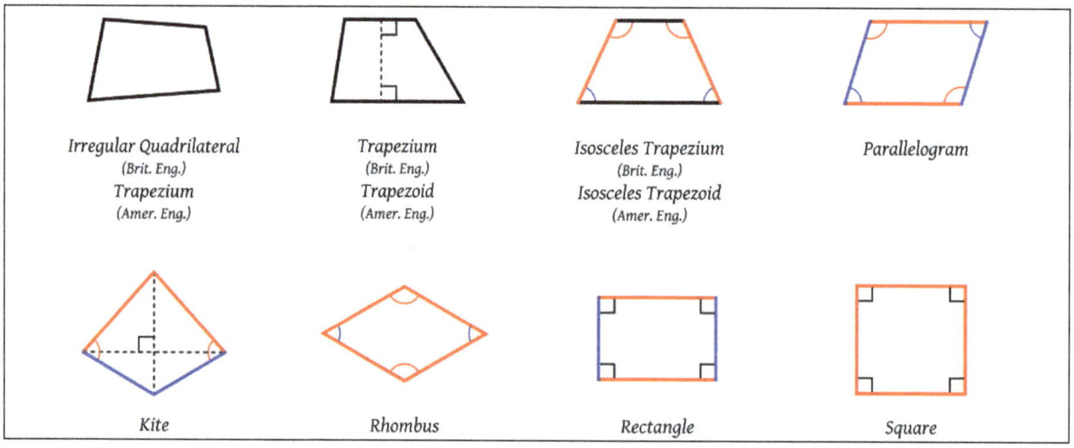

- Tangential quadrilateral: the four sides are tangents to an inscribed circle. A convex quadrilateral is tangential if and only if opposite sides have equal sums.
- Tangential trapezoid: a trapezoid where the four sides are tangents to an inscribed circle.
- Cyclic quadrilateral: the four vertices lie on a circumscribed circle. A convex quadrilateral is cyclic if and only if opposite angles sum to 180°.
- Right kite: a kite with two opposite right angles. It is a type of cyclic quadrilateral.
- Bicentric quadrilateral: it is both tangential and cyclic.
- Orthodiagonal quadrilateral: the diagonals cross at right angles.
- Equidiagonal quadrilateral: the diagonals are of equal length.
- Ex-tangential quadrilateral: the four extensions of the sides are tangent to an excircle.
- An *equilic quadrilateral* has two opposite equal sides that, when extended, meet at 60°.
- A *Watt quadrilateral* is a quadrilateral with a pair of opposite sides of equal length.
- A *quadric quadrilateral* is a convex quadrilateral whose four vertices all lie on the perimeter of a square.

- A *diametric quadrilateral* is a cyclic quadrilateral having one of its sides as a diameter of the circumcircle.

Concave Quadrilaterals

In a concave quadrilateral, one interior angle is bigger than 180° and one of the two diagonals lies outside the quadrilateral.

- A *dart* (or arrowhead) is a concave quadrilateral with bilateral symmetry like a kite, but one interior angle is reflex.

Complex Quadrilaterals

A self-intersecting quadrilateral is called variously a cross-quadrilateral, crossed quadrilateral, butterfly quadrilateral or bow-tie quadrilateral. In a crossed quadrilateral, the four "interior" angles on either side of the crossing (two acute and two reflex, all on the left or all on the right as the figure is traced out) add up to 720°.

An antiparallelogram

- Antiparallelogram: a crossed quadrilaterals in which (like a parallelogram) each pair of nonadjacent sides have equal lengths.
- Crossed rectangle: an antiparallelogram whose sides are two opposite sides and the two diagonals of a rectangle, hence having one pair of opposite sides parallel.
- Crossed square: a special case of a crossed rectangle where two of the sides intersect at right angles.

Special Line Segments

The two diagonals of a convex quadrilateral are the line segments that connect opposite vertices.

The two bimedians of a convex quadrilateral are the line segments that connect the midpoints of opposite sides. They intersect at the "vertex centroid" of the quadrilateral.

The four maltitudes of a convex quadrilateral are the perpendiculars to a side through the midpoint of the opposite side.

Area of a Convex Quadrilateral

There are various general formulas for the area K of a convex quadrilateral $ABCD$ with sides $a = AB$, $b = BC$, $c = CD$ and $d = DA$.

Trigonometric Formulas

The area can be expressed in trigonometric terms as,

$$K = \tfrac{1}{2} pq \cdot \sin\theta,$$

where the lengths of the diagonals are p and q and the angle between them is θ. In the case of an orthodiagonal quadrilateral (e.g. rhombus, square, and kite), this formula reduces to $K = \tfrac{1}{2} pq$ since θ is 90°.

The area can be also expressed in terms of bimedians as,

$$K = mn \cdot \sin\varphi,$$

where the lengths of the bimedians are m and n and the angle between them is φ.

Bretschneider's formula expresses the area in terms of the sides and two opposite angles:

$$\begin{aligned} K &= \sqrt{(s-a)(s-b)(s-c)(s-d) - \frac{1}{2} abcd\,[1+\cos(A+C)]} \\ &= \sqrt{(s-a)(s-b)(s-c)(s-d) - abcd[\cos^2(\frac{A+C}{2})]} \end{aligned}$$

where the sides in sequence are a, b, c, d, where s is the semiperimeter, and A and C are two (in fact, any two) opposite angles. This reduces to Brahmagupta's formula for the area of a cyclic quadrilateral when $A+C = 180°$.

Another area formula in terms of the sides and angles, with angle C being between sides b and c, and A being between sides a and d, is:

$$K = \tfrac{1}{2} ad \cdot \sin A + \tfrac{1}{2} bc \cdot \sin C.$$

In the case of a cyclic quadrilateral, the latter formula becomes $K = \frac{1}{2}(ad + bc)\sin A$.

In a parallelogram, where both pairs of opposite sides and angles are equal, this formula reduces to $K = ab \cdot \sin A$.

Alternatively, we can write the area in terms of the sides and the intersection angle θ of the diagonals, so long as this angle is not 90°:

$$K = \frac{|\tan\theta|}{4} \cdot \left|a^2 + c^2 - b^2 - d^2\right|.$$

In the case of a parallelogram, the latter formula becomes $K = \frac{1}{2} |\tan\theta| \cdot |a^2 - b^2|$.

Another area formula including the sides a, b, c, d is:

$$K = \tfrac{1}{4}\sqrt{(2(a^2+c^2)-4x^2)(2(b^2+d^2)-4x^2)}\sin\varphi$$

where x is the distance between the midpoints of the diagonals and φ is the angle between the bimedians.

The last trigonometric area formula including the sides a, b, c, d and the angle α between a and b is:

$$K = \frac{1}{2}ab\cdot\sin\alpha + \frac{1}{4}\sqrt{4c^2d^2-(c^2+d^2-a^2-b^2+2ab\cdot\cos\alpha)^2},$$

which can also be used for the area of a concave quadrilateral (having the concave part opposite to angle α) just changing the first sign + to - .

Non-trigonometric Formulas

The following two formulas express the area in terms of the sides a, b, c, d, the semiperimeter s, and the diagonals p, q:

$$K = \sqrt{(s-a)(s-b)(s-c)(s-d)-\tfrac{1}{4}(ac+bd+pq)(ac+bd-pq)},$$

$$K = \tfrac{1}{4}\sqrt{4p^2q^2-\left(a^2+c^2-b^2-d^2\right)^2}.$$

The first reduces to Brahmagupta's formula in the cyclic quadrilateral case, since then $pq = ac + bd$.

The area can also be expressed in terms of the bimedians m, n and the diagonals p, q:

$$K = \tfrac{1}{2}\sqrt{(m+n+p)(m+n-p)(m+n+q)(m+n-q)},$$

$$K = \tfrac{1}{2}\sqrt{p^2q^2-(m^2-n^2)^2}.$$

In fact, any three of the four values m, n, p, and q suffice for determination of the area, since in any quadrilateral the four values are related by $p^2+q^2=2(m^2+n^2)$. The corresponding expressions are:

$$K = \tfrac{1}{2}\sqrt{[(m+n)^2-p^2]\cdot[p^2-(m-n)^2]},$$

if the lengths of two bimedians and one diagonal are given,

$$K = \tfrac{1}{2}\sqrt{[(m+n)^2-p^2]\cdot[p^2-(m-n)^2]},$$

if the lengths of two diagonals and one bimedian are given.

Vector Formulas

The area of a quadrilateral *ABCD* can be calculated using vectors. Let vectors AC and BD form the diagonals from *A* to *C* and from *B* to *D*. The area of the quadrilateral is then,

$$K = \tfrac{1}{2}|\mathbf{AC} \times \mathbf{BD}|,$$

which is half the magnitude of the cross product of vectors AC and BD. In two-dimensional Euclidean space, expressing vector AC as a free vector in Cartesian space equal to $(\boldsymbol{x}_1,\boldsymbol{y}_1)$ and BD as $(\boldsymbol{x}_2,\boldsymbol{y}_2)$, this can be rewritten as:

$$K = \tfrac{1}{2}|x_1 y_2 - x_2 y_1|.$$

Diagonals

Properties of the Diagonals in Some Quadrilaterals

In the following table it is listed if the diagonals in some of the most basic quadrilaterals bisect each other, if their diagonals are perpendicular, and if their diagonals have equal length. The list applies to the most general cases, and excludes named subsets.

Quadrilateral	Bisecting diagonals	Perpendicular diagonals	Equal diagonals
Trapezoid	No	*See note 1*	No
Isosceles trapezoid	No	*See note 1*	Yes
Parallelogram	Yes	No	No
Kite	*See note 2*	Yes	*See note 2*
Rectangle	Yes	No	Yes
Rhombus	Yes	Yes	No
Square	Yes	Yes	Yes

Note 1: The most general trapezoids and isosceles trapezoids do not have perpendicular diagonals, but there are infinite numbers of (non-similar) trapezoids and isosceles trapezoids that do have perpendicular diagonals and are not any other named quadrilateral.

Note 2: In a kite, one diagonal bisects the other. The most general kite has unequal diagonals, but there is an infinite number of (non-similar) kites in which the diagonals are equal in length (and the kites are not any other named quadrilateral).

Lengths of the Diagonals

The lengths of the diagonals in a convex quadrilateral *ABCD* can be calculated using the law of cosines on each triangle formed by one diagonal and two sides of the quadrilateral. Thus,

$$p = \sqrt{a^2 + b^2 - 2ab\cos B} = \sqrt{c^2 + d^2 - 2cd\cos D}$$

and

$$q = \sqrt{a^2 + d^2 - 2ad\cos A} = \sqrt{b^2 + c^2 - 2bc\cos C}.$$

Other, more symmetric formulas for the lengths of the diagonals, are

$$p = \sqrt{\frac{(ac+bd)(ad+bc)-2abcd(\cos B+\cos D)}{ab+cd}}$$

and

$$q = \sqrt{\frac{(ab+cd)(ac+bd)-2abcd(\cos A+\cos C)}{ad+bc}}.$$

Generalizations of the Parallelogram Law and Ptolemy's Theorem

In any convex quadrilateral *ABCD*, the sum of the squares of the four sides is equal to the sum of the squares of the two diagonals plus four times the square of the line segment connecting the midpoints of the diagonals. Thus,

$$a^2+b^2+c^2+d^2=p^2+q^2+4x^2$$

where x is the distance between the midpoints of the diagonals. This is sometimes known as *Euler's quadrilateral theorem* and is a generalization of the parallelogram law.

The German mathematician Carl Anton Bretschneider derived in 1842 the following generalization of Ptolemy's theorem, regarding the product of the diagonals in a convex quadrilateral

$$p^2q^2=a^2c^2+b^2d^2-2abcd\cos(A+C).$$

This relation can be considered to be a law of cosines for a quadrilateral. In a cyclic quadrilateral, where $A + C = 180°$, it reduces to $pq = ac + bd$. Since $\cos(A + C) \geq -1$, it also gives a proof of Ptolemy's inequality.

Other Metric Relations

If X and Y are the feet of the normals from B and D to the diagonal $AC = p$ in a convex quadrilateral *ABCD* with sides $a = AB$, $b = BC$, $c = CD$, $d = DA$, then,

$$XY=\frac{|a^2+c^2-b^2-d^2|}{2p}.$$

In a convex quadrilateral *ABCD* with sides $a = AB$, $b = BC$, $c = CD$, $d = DA$, and where the diagonals intersect at E,

$$efgh(a+c+b+d)(a+c-b-d)=(agh+cef+beh+dfg)(agh+cef-beh-dfg)$$

where, $e = AE$, $f = BE$, $g = CE$, and $h = DE$.

The shape and size of a convex quadrilateral are fully determined by the lengths of its sides in sequence and of one diagonal between two specified vertices. The two diagonals p, q and the four side lengths a, b, c, d of a quadrilateral are related by the Cayley-Menger determinant, as follows:

$$\det\begin{bmatrix} 0 & a^2 & p^2 & d^2 & 1 \\ a^2 & 0 & b^2 & q^2 & 1 \\ p^2 & b^2 & 0 & c^2 & 1 \\ d^2 & q^2 & c^2 & 0 & 1 \\ 1 & 1 & 1 & 1 & 0 \end{bmatrix} = 0.$$

Angle Bisectors

The internal angle bisectors of a convex quadrilateral either form a cyclic quadrilateral (that is, the four intersection points of adjacent angle bisectors are concyclic) or they are concurrent. In the latter case the quadrilateral is a tangential quadrilateral.

In quadrilateral *ABCD*, if the angle bisectors of *A* and *C* meet on diagonal *BD*, then the angle bisectors of *B* and *D* meet on diagonal *AC*.

Bimedians

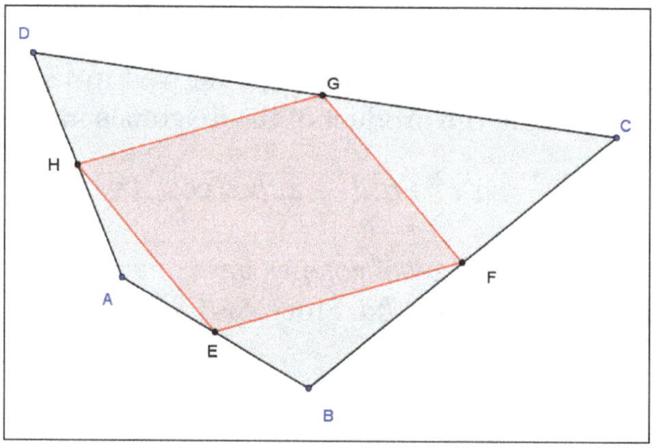

The Varignon parallelogram *EFGH*.

The bimedians of a quadrilateral are the line segments connecting the midpoints of the opposite sides. The intersection of the bimedians is the centroid of the vertices of the quadrilateral.

The midpoints of the sides of any quadrilateral (convex, concave or crossed) are the vertices of a parallelogram called the Varignon parallelogram. It has the following properties:

- Each pair of opposite sides of the Varignon parallelogram are parallel to a diagonal in the original quadrilateral.
- A side of the Varignon parallelogram is half as long as the diagonal in the original quadrilateral it is parallel to.
- The area of the Varignon parallelogram equals half the area of the original quadrilateral. This is true in convex, concave and crossed quadrilaterals provided the area of the latter is defined to be the difference of the areas of the two triangles it is composed of.

- The perimeter of the Varignon parallelogram equals the sum of the diagonals of the original quadrilateral.
- The diagonals of the Varignon parallelogram are the bimedians of the original quadrilateral.

The two bimedians in a quadrilateral and the line segment joining the midpoints of the diagonals in that quadrilateral are concurrent and are all bisected by their point of intersection.

In a convex quadrilateral with sides a, b, c and d, the length of the bimedian that connects the midpoints of the sides a and c is:

$$m = \tfrac{1}{2}\sqrt{-a^2 + b^2 - c^2 + d^2 + p^2 + q^2}$$

where p and q are the length of the diagonals. The length of the bimedian that connects the midpoints of the sides b and d is:

$$n = \tfrac{1}{2}\sqrt{a^2 - b^2 + c^2 - d^2 + p^2 + q^2}.$$

Hence

$$p^2 + q^2 = 2(m^2 + n^2).$$

This is also a corollary to the parallelogram law applied in the Varignon parallelogram.

The lengths of the bimedians can also be expressed in terms of two opposite sides and the distance x between the midpoints of the diagonals. This is possible when using Euler's quadrilateral theorem in the above formulas. Whence,

$$m = \tfrac{1}{2}\sqrt{2(b^2 + d^2) - 4x^2}$$

and

$$n = \tfrac{1}{2}\sqrt{2(a^2 + c^2) - 4x^2}.$$

Note that the two opposite sides in these formulas are not the two that the bimedian connects.

In a convex quadrilateral, there is the following dual connection between the bimedians and the diagonals:

- The two bimedians have equal length if and only if the two diagonals are perpendicular.
- The two bimedians are perpendicular if and only if the two diagonals have equal length.

Trigonometric identities

The four angles of a simple quadrilateral $ABCD$ satisfy the following identities:

$$A+\sin B+\sin C+\sin D=4\sin\frac{A+B}{2}\sin\frac{A+C}{2}\sin\frac{A+D}{2}$$

and

$$\frac{\tan A\tan B-\tan C\tan D}{\tan A\tan C-\tan B\tan D}=\frac{\tan(A+C)}{\tan(A+B)}.$$

Also,

$$\frac{A+\tan B+\tan C+\tan D}{\cot A+\cot B+\cot C+\cot D}=\tan A\tan B\tan C\tan D.$$

In the last two formulas, no angle is allowed to be a right angle, since tan 90° is not defined.

Inequalities

Area

If a convex quadrilateral has the consecutive sides a, b, c, d and the diagonals p, q, then its area K satisfies,

$K\le\frac{1}{4}(a+c)(b+d)$ with equality only for a rectangle.

$K\le\frac{1}{4}(a^2+b^2+c^2+d^2)$ with equality only for a square.

$K\le\frac{1}{4}(p^2+q^2)$ with equality only if the diagonals are perpendicular and equal.

$K\le\frac{1}{2}\sqrt{(a^2+c^2)(b^2+d^2)}$ with equality only for a rectangle.

From Bretschneider's formula it directly follows that the area of a quadrilateral satisfies,

$$K\le\sqrt{(s-a)(s-b)(s-c)(s-d)}$$

with equality if and only if the quadrilateral is cyclic or degenerate such that one side is equal to the sum of the other three (it has collapsed into a line segment, so the area is zero).

The area of any quadrilateral also satisfies the inequality,

$$K\le\tfrac{1}{2}\sqrt[3]{(ab+cd)(ac+bd)(ad+bc)}.$$

Denoting the perimeter as L, we have,

$$K\le\tfrac{1}{16}L^2,$$

with equality only in the case of a square.

The area of a convex quadrilateral also satisfies,

$$K\le\tfrac{1}{2}pq$$

for diagonal lengths p and q, with equality if and only if the diagonals are perpendicular.

Diagonals and Bimedians

A corollary to Euler's quadrilateral theorem is the inequality,

$$a^2 + b^2 + c^2 + d^2 \geq p^2 + q^2$$

where equality holds if and only if the quadrilateral is a parallelogram.

Euler also generalized Ptolemy's theorem, which is an equality in a cyclic quadrilateral, into an inequality for a convex quadrilateral. It states that,

$$pq \leq ac + bd$$

where there is equality if and only if the quadrilateral is cyclic. This is often called Ptolemy's inequality.

In any convex quadrilateral the bimedians m, n and the diagonals p q are related by the inequality,

$$pq \leq m^2 + n^2,$$

with equality holding if and only if the diagonals are equal. This follows directly from the quadrilateral identity $m^2 + n^2 = \tfrac{1}{2}(p^2 + q^2)$.

Sides

The sides a, b, c, and d of any quadrilateral satisfy,

$$a^2 + b^2 + c^2 > \frac{d^2}{3}$$

and

$$a^4 + b^4 + c^4 \geq \frac{d^4}{27}.$$

Maximum and Minimum Properties

Among all quadrilaterals with a given perimeter, the one with the largest area is the square. This is called the *isoperimetric theorem for quadrilaterals*. It is a direct consequence of the area inequality,

$$K \leq \tfrac{1}{16} L^2$$

where, K is the area of a convex quadrilateral with perimeter L. Equality holds if and only if the quadrilateral is a square. The dual theorem states that of all quadrilaterals with a given area, the square has the shortest perimeter.

The quadrilateral with given side lengths that has the maximum area is the cyclic quadrilateral.

Of all convex quadrilaterals with given diagonals, the orthodiagonal quadrilateral has the largest area. This is a direct consequence of the fact that the area of a convex quadrilateral satisfies,

$$K = \tfrac{1}{2} pq \sin\theta \le \tfrac{1}{2} pq,$$

where θ is the angle between the diagonals p and q. Equality holds if and only if $\theta = 90°$.

If P is an interior point in a convex quadrilateral $ABCD$, then,

$$AP + BP + CP + DP \ge AC + BD.$$

From this inequality it follows that the point inside a quadrilateral that minimizes the sum of distances to the vertices is the intersection of the diagonals. Hence that point is the Fermat point of a convex quadrilateral.

Remarkable Points and Lines in a Convex Quadrilateral

The centre of a quadrilateral can be defined in several different ways. The "vertex centroid" comes from considering the quadrilateral as being empty but having equal masses at its vertices. The "side centroid" comes from considering the sides to have constant mass per unit length. The usual centre, called just centroid (centre of area) comes from considering the surface of the quadrilateral as having constant density. These three points are in general not all the same point.

The "vertex centroid" is the intersection of the two bimedians. As with any polygon, the x and y coordinates of the vertex centroid are the arithmetic means of the x and y coordinates of the vertices.

The "area centroid" of quadrilateral $ABCD$ can be constructed in the following way. Let G_a, G_b, G_c, G_d be the centroids of triangles BCD, ACD, ABD, ABC respectively. Then the "area centroid" is the intersection of the lines G_aG_c and G_bG_d.

In a general convex quadrilateral $ABCD$, there are no natural analogies to the circumcenter and orthocenter of a triangle. But two such points can be constructed in the following way. Let O_a, O_b, O_c, O_d be the circumcenters of triangles BCD, ACD, ABD, ABC respectively; and denote by H_a, H_b, H_c, H_d the orthocenters in the same triangles. Then the intersection of the lines O_aO_c and O_bO_d is called the quasicircumcenter, and the intersection of the lines H_aH_c and H_bH_d is called the *quasiorthocenter* of the convex quadrilateral. These points can be used to define an Euler line of a quadrilateral. In a convex quadrilateral, the quasiorthocenter H, the "area centroid" G, and the quasicircumcenter O are collinear in this order, and $HG = 2GO$.

There can also be defined a *quasinine-point center* E as the intersection of the lines E_aE_c and E_bE_d, where E_a, E_b, E_c, E_d are the nine-point centers of triangles BCD, ACD, ABD, ABC respectively. Then E is the midpoint of OH.

Another remarkable line in a convex non-parallelogram quadrilateral is the Newton line, which connects the midpoints of the diagonals, and which contains the intersection of the bimedians.

Other Properties of Convex Quadrilaterals

- Let exterior squares be drawn on all sides of a quadrilateral. The segments connecting the centers of opposite squares are (a) equal in length, and (b) perpendicular. Thus these centers are the vertices of an orthodiagonal quadrilateral. This is called Van Aubel's theorem.
- For any simple quadrilateral with given edge lengths, there is a cyclic quadrilateral with the same edge lengths.
- The four smaller triangles formed by the diagonals and sides of a convex quadrilateral have the property that the product of the areas of two opposite triangles equals the product of the areas of the other two triangles.

Taxonomy

A hierarchical taxonomy of quadrilaterals is illustrated by the figure to the right. Lower classes are special cases of higher classes they are connected to. Note that "trapezoid" here is referring to the North American definition (the British equivalent is a trapezium). Inclusive definitions are used throughout.

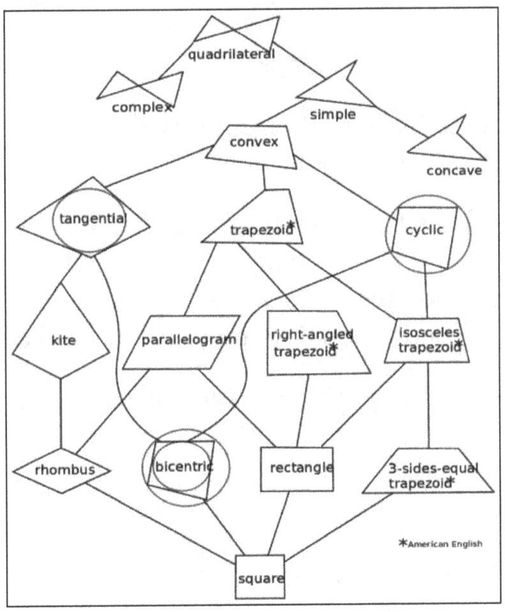

A taxonomy of quadrilaterals.

Skew Quadrilaterals

A non-planar quadrilateral is called a skew quadrilateral. Formulas to compute its dihedral angles from the edge lengths and the angle between two adjacent edges were derived for work on the properties of molecules such as cyclobutane that contain a "puckered" ring of four atoms. Skew polygon for more. Historically the term gauche quadrilateral was also used to mean a skew quadrilateral. A skew quadrilateral together with its diagonals form a (possibly non-regular) tetrahedron, and conversely every skew quadrilateral comes from a tetrahedron where a pair of opposite edges is removed.

Rectangle

In Euclidean plane geometry, a rectangle is a quadrilateral with four right angles. It can also be defined as an equiangular quadrilateral, since equiangular means that all of its angles are equal ($360°/4 = 90°$). It can also be defined as a parallelogram containing a right angle. A rectangle with four sides of equal length is a square. The term oblong is occasionally used to refer to a non-square rectangle. A rectangle with vertices *ABCD* would be denoted as ▭*ABCD*.

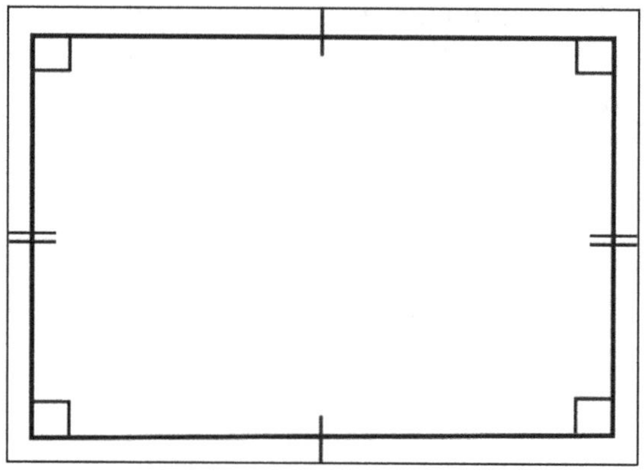

Rectangle

The word rectangle comes from the Latin *rectangulus*, which is a combination of *rectus* (right) and *angulus* (angle).

A crossed rectangle is a crossed (self-intersecting) quadrilateral which consists of two opposite sides of a rectangle along with the two diagonals. It is a special case of an antiparallelogram, and its angles are not right angles. Other geometries, such as spherical, elliptic, and hyperbolic, have so-called rectangles with opposite sides equal in length and equal angles that are not right angles.

Rectangles are involved in many tiling problems, such as tiling the plane by rectangles or tiling a rectangle by polygons.

Characterizations

A convex quadrilateral is a rectangle if and only if it is any one of the following:

- An equiangular quadrilateral
- A quadrilateral with four right angles
- A parallelogram with at least one right angle
- A parallelogram with diagonals of equal length
- A parallelogram *ABCD* where triangles *ABD* and *DCA* are congruent
- A convex quadrilateral with successive sides *a*, *b*, *c*, *d* whose area is $\frac{1}{4}(a+c)(b+d)$.
- A convex quadrilateral with successive sides *a*, *b*, *c*, *d* whose area is $\frac{1}{2}\sqrt{(a^2+c^2)(b^2+d^2)}$.

Classification

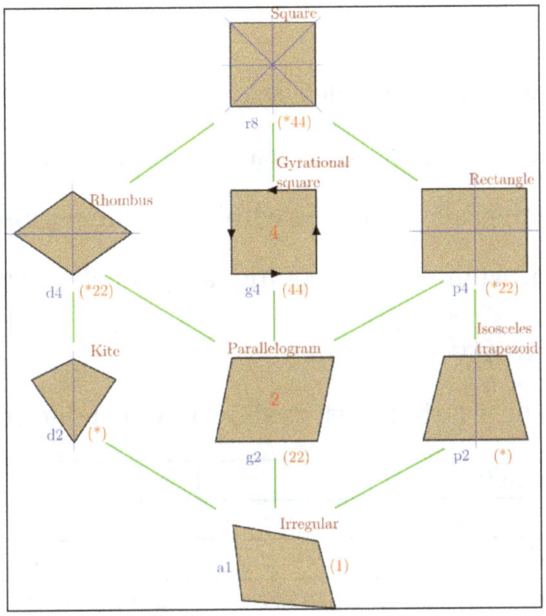

A rectangle is a special case of both parallelogram and trapezoid. A square is a special case of a rectangle.

Traditional Hierarchy

A rectangle is a special case of a parallelogram in which each pair of adjacent sides is perpendicular.

A parallelogram is a special case of a trapezium (known as a trapezoid in North America) in which *both* pairs of opposite sides are parallel and equal in length.

A trapezium is a convex quadrilateral which has at least one pair of parallel opposite sides.

A convex quadrilateral is

- Simple: The boundary does not cross itself.
- Star-shaped: The whole interior is visible from a single point, without crossing any edge.

Alternative Hierarchy

De Villiers defines a rectangle more generally as any quadrilateral with axes of symmetry through each pair of opposite sides. This definition includes both right-angled rectangles and crossed rectangles. Each has an axis of symmetry parallel to and equidistant from a pair of opposite sides, and another which is the perpendicular bisector of those sides, but, in the case of the crossed rectangle, the first axis is not an axis of symmetry for either side that it bisects.

Quadrilaterals with two axes of symmetry, each through a pair of opposite sides, belong to the larger class of quadrilaterals with at least one axis of symmetry through a pair of opposite sides. These quadrilaterals comprise isosceles trapezia and crossed isosceles trapezia (crossed quadrilaterals with the same vertex arrangement as isosceles trapezia).

Properties

Symmetry

A rectangle is cyclic: all corners lie on a single circle.

It is equiangular: all its corner angles are equal (each of 90 degrees).

It is isogonal or vertex-transitive: all corners lie within the same symmetry orbit.

It has two lines of reflectional symmetry and rotational symmetry of order 2 (through 180°).

Rectangle-rhombus Duality

The dual polygon of a rectangle is a rhombus, as shown in the table below.

Rectangle	Rhombus
All *angles* are equal.	All *sides* are equal.
Alternate *sides* are equal.	Alternate *angles* are equal.
Its centre is equidistant from its *vertices*, hence it has a *circumcircle*.	Its centre is equidistant from its *sides*, hence it has an *incircle*.
Its axes of symmetry bisect opposite *sides*.	Its axes of symmetry bisect opposite *angles*.
Diagonals are equal in *length*.	Diagonals intersect at equal *angles*.

- The figure formed by joining, in order, the midpoints of the sides of a rectangle is a rhombus and vice versa.

Miscellaneous

The two diagonals are equal in length and bisect each other. Every quadrilateral with both these properties is a rectangle.

A rectangle is rectilinear: its sides meet at right angles.

A rectangle in the plane can be defined by five independent degrees of freedom consisting, for example, of three for position (comprising two of translation and one of rotation), one for shape (aspect ratio), and one for overall size (area).

Two rectangles, neither of which will fit inside the other, are said to be incomparable.

Formulae

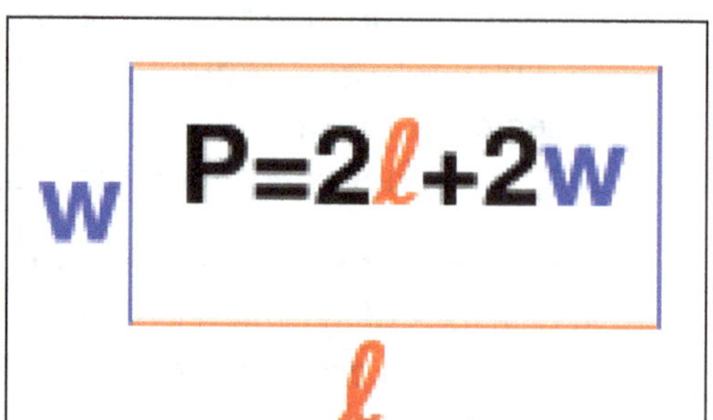

The formula for the perimeter of a rectangle.

If a rectangle has length ℓ and width w,

- It has area $A = \ell w$,
- It has perimeter $P = 2\ell + 2w = 2(\ell + w)$,
- Each diagonal has length $d = \sqrt{\ell^2 + w^2}$,
- When $\ell = w$, the rectangle is a square.

Theorems

The isoperimetric theorem for rectangles states that among all rectangles of a given perimeter, the square has the largest area.

The midpoints of the sides of any quadrilateral with perpendicular diagonals form a rectangle.

A parallelogram with equal diagonals is a rectangle.

The Japanese theorem for cyclic quadrilaterals states that the incentres of the four triangles determined by the vertices of a cyclic quadrilateral taken three at a time form a rectangle.

The British flag theorem states that with vertices denoted *A*, *B*, *C*, and *D*, for any point *P* on the same plane of a rectangle:

$$(AP)^2 + (CP)^2 = (BP)^2 + (DP)^2.$$

For every convex body C in the plane, we can inscribe a rectangle r in C such that a homothetic copy R of r is circumscribed about C and the positive homothety ratio is at most 2 and $0.5 \times \text{Area}(R) \leq \text{Area}(C) \leq 2 \times \text{Area}(r)$.

Crossed Rectangles

A crossed (self-intersecting) quadrilateral consists of two opposite sides of a non-self-intersecting quadrilateral along with the two diagonals. Similarly, a crossed rectangle is a crossed quadrilateral which consists of two opposite sides of a rectangle along with the two diagonals. It has the same vertex arrangement as the rectangle. It appears as two identical triangles with a common vertex, but the geometric intersection is not considered a vertex.

A crossed quadrilateral is sometimes likened to a bow tie or butterfly. A three-dimensional rectangular wire frame that is twisted can take the shape of a bow tie. A crossed rectangle is sometimes called an "angular eight".

The interior of a crossed rectangle can have a polygon density of ±1 in each triangle, dependent upon the winding orientation as clockwise or counterclockwise.

A crossed rectangle is not equiangular. The sum of its interior angles (two acute and two reflex), as with any crossed quadrilateral, is 720°.

A rectangle and a crossed rectangle are quadrilaterals with the following properties in common:

- Opposite sides are equal in length.
- The two diagonals are equal in length.
- It has two lines of reflectional symmetry and rotational symmetry of order 2 (through 180°).

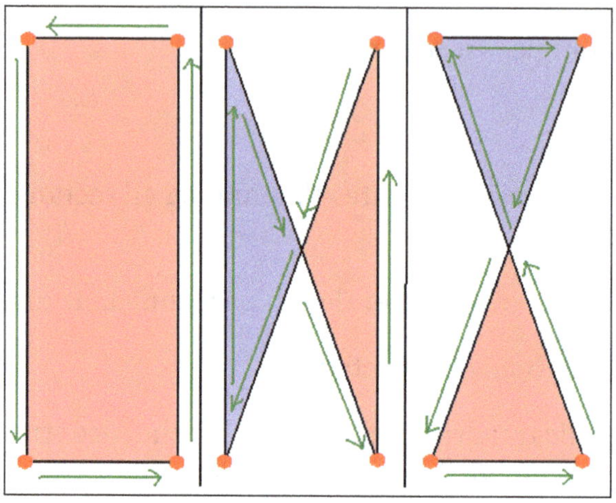

Other Rectangles

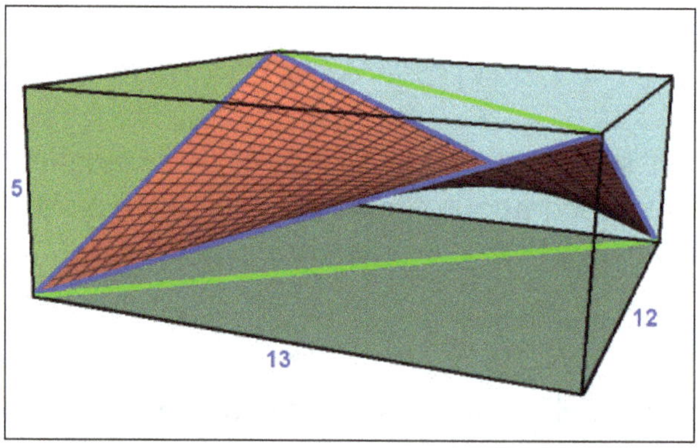

A saddle rectangle has 4 nonplanar vertices, alternated from vertices of a cuboid, with a unique minimal surface interior defined as a linear combination of the four vertices, creating a saddle surface. This example shows 4 blue edges of the rectangle, and two green diagonals, all being diagonal of the cuboid rectangular faces.

In spherical geometry, a spherical rectangle is a figure whose four edges are great circle arcs which meet at equal angles greater than 90°. Opposite arcs are equal in length. The surface of a sphere in Euclidean solid geometry is a non-Euclidean surface in the sense of elliptic geometry. Spherical geometry is the simplest form of elliptic geometry.

In elliptic geometry, an elliptic rectangle is a figure in the elliptic plane whose four edges are elliptic arcs which meet at equal angles greater than 90°. Opposite arcs are equal in length.

In hyperbolic geometry, a hyperbolic rectangle is a figure in the hyperbolic plane whose four edges are hyperbolic arcs which meet at equal angles less than 90°. Opposite arcs are equal in length.

Tessellations

The rectangle is used in many periodic tessellation patterns, in brickwork, for example, these tilings:

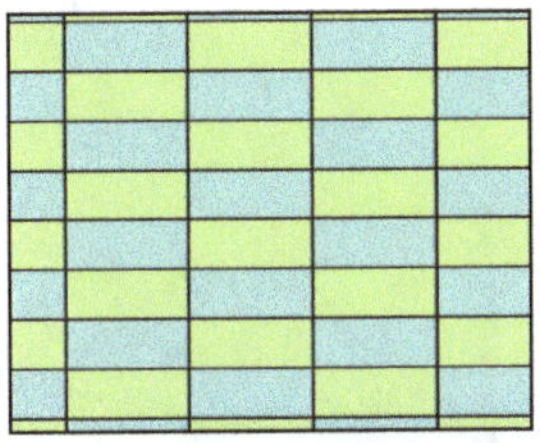
Stacked bond

Running bond

Basket weave

Basket weave

Herringbone pattern

Squared, Perfect, and Other Tiled Rectangles

A rectangle tiled by squares, rectangles, or triangles is said to be a “squared”, “rectangled”, or “triangulated” (or “triangled”) rectangle respectively. The tiled rectangle is *perfect* if the tiles are similar and finite in number and no two tiles are the same size. If two such tiles are the same size, the tiling is *imperfect*. In a perfect (or imperfect) triangled rectangle the triangles must be right triangles.

A rectangle has commensurable sides if and only if it is tileable by a finite number of unequal squares. The same is true if the tiles are unequal isosceles right triangles.

The tilings of rectangles by other tiles which have attracted the most attention are those by congruent non-rectangular polyominoes, allowing all rotations and reflections. There are also tilings by congruent polyaboloes.

Square

In geometry, a square is a regular quadrilateral, which means that it has four equal sides and four equal angles (90-degree angles, or right angles). It can also be defined as a rectangle in which two adjacent sides have equal length. A square with vertices *ABCD* would be denoted □*ABCD*.

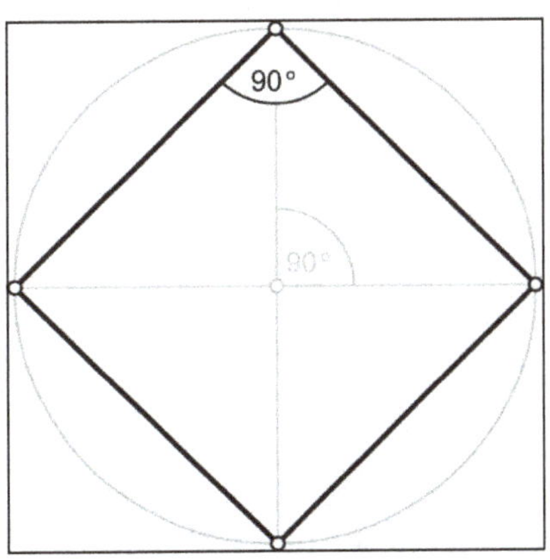

A regular quadrilateral (tetragon).

Properties

A square is a special case of a rhombus (equal sides, opposite equal angles), a kite (two pairs of adjacent equal sides), a parallelogram (opposite sides parallel), a quadrilateral or tetragon (four-sided polygon), and a rectangle (opposite sides equal, right-angles) and therefore has all the properties of all these shapes, namely:

- The diagonals of a square bisect each other and meet at 90°.
- The diagonals of a square bisect its angles.
- Opposite sides of a square are both parallel and equal in length.
- All four angles of a square are equal. (Each is 360°/4 = 90°, so every angle of a square is a right angle.)
- All four sides of a square are equal.
- The diagonals of a square are equal.
- The square is the n=2 case of the families of n-hypercubes and n-orthoplexes.
- A square has Schläfli symbol {4}. A truncated square, t{4}, is an octagon, {8}. An alternated square, h{4}, is a digon, {2}.

Perimeter and Area

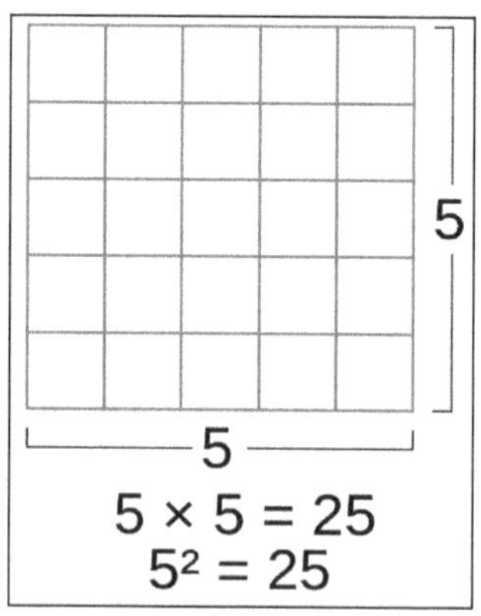

The area of a square is the product of the length of its sides.

The perimeter of a square whose four sides have length ℓ is,

$$P = 4\ell$$

and the area A is

In classical times, the second power was described in terms of the area of a square, as in the above formula. This led to the use of the term *square* to mean raising to the second power.

The area can also be calculated using the diagonal d according to,

$$A = \frac{d^2}{2}.$$

In terms of the circumradius R, the area of a square is,

$$A = 2R^2;$$

since the area of the circle is πR^2, the square fills approximately 0.6366 of its circumscribed circle.

In terms of the inradius r, the area of the square is,

$$A = 4r^2.$$

Because it is a regular polygon, a square is the quadrilateral of least perimeter enclosing a given area. Dually, a square is the quadrilateral containing the largest area within a given perimeter. Indeed, if A and P are the area and perimeter enclosed by a quadrilateral, then the following isoperimetric inequality holds:

$$16A \le P^2$$

with equality if and only if the quadrilateral is a square.

A convex quadrilateral with successive sides a, b, c, d is a square if and only if,

$$A = \frac{1}{2}(a^2 + c^2) = \frac{1}{2}(b^2 + d^2).$$

Other Facts

- The diagonals of a square are $\sqrt{2}$ (about 1.414) times the length of a side of the square. This value, known as the square root of 2 or Pythagoras' constant, was the first number proven to be irrational.
- A square can also be defined as a parallelogram with equal diagonals that bisect the angles.
- If a figure is both a rectangle (right angles) and a rhombus (equal edge lengths), then it is a square.
- If a circle is circumscribed around a square, the area of the circle is $\pi / 2$ (about 1.5708) times the area of the square.
- If a circle is inscribed in the square, the area of the circle is $\pi / 4$ (about 0.7854) times the area of the square.
- A square has a larger area than any other quadrilateral with the same perimeter.
- A square tiling is one of three regular tilings of the plane (the others are the equilateral triangle and the regular hexagon).

- The square is in two families of polytopes in two dimensions: hypercube and the cross polytope. The Schläfli symbol for the square is {4}.
- The square is a highly symmetric object. There are four lines of reflectional symmetry and it has rotational symmetry of order 4 (through 90°, 180° and 270°). Its symmetry group is the dihedral group D_4.
- If the inscribed circle of a square *ABCD* has tangency points *E* on *AB*, *F* on *BC*, *G* on *CD*, and *H* on *DA*, then for any point *P* on the inscribed circle,

  $$2(PH^2 - PE^2) = PD^2 - PB^2.$$

- If d_i is the distance from an arbitrary point in the plane to the *i*-th vertex of a square and R is the circumradius of the square, then

  $$\tfrac{1}{4}(d_1^4 + d_2^4 + d_3^4 + d_4^4) + 3R^4 = \left(\frac{d_1^2 + d_2^2 + d_3^2 + d_4^2}{4} + R^2\right)^2.$$

Coordinates and Equations

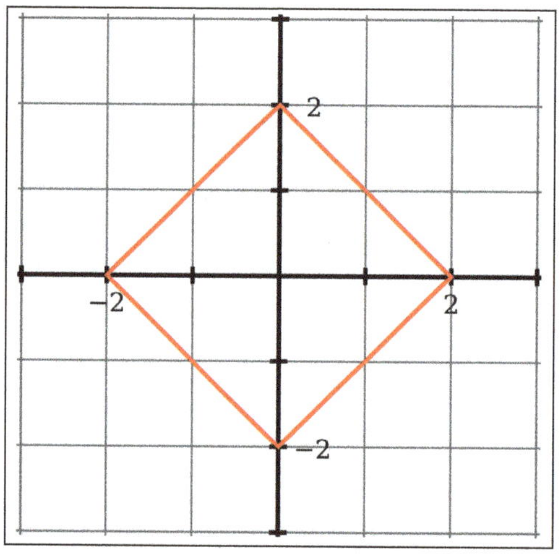

plotted on *Cartesian coordinates.*

The coordinates for the vertices of a square with vertical and horizontal sides, centered at the origin and with side length 2 are (±1, ±1), while the interior of this square consists of all points (x_i, y_i) with $-1 < x_i < 1$ and $-1 < y_i < 1$. The equation,

$$\max(x^2, y^2) = 1$$

specifies the boundary of this square. This equation means "x^2 or y^2, whichever is larger, equals 1." The circumradius of this square (the radius of a circle drawn through the square's vertices) is half the square's diagonal, and equals $\sqrt{2}$. Then the circumcircle has the equation,

$$x^2 + y^2 = 2.$$

Alternatively the equation,

$$|x - a| + |y - b| = r.$$

can also be used to describe the boundary of a square with center coordinates (a, b) and a horizontal or vertical radius of r.

Construction

The following animations show how to construct a square using a compass and straightedge. This is possible as $4 = 2^2$, a power of two.

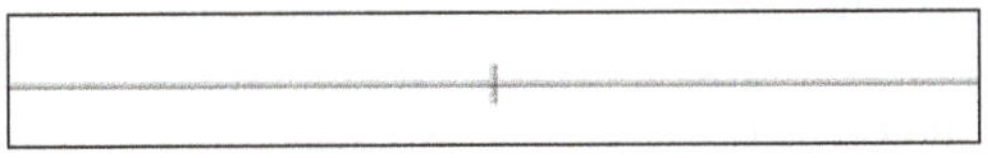

Square at a given circumcircle.

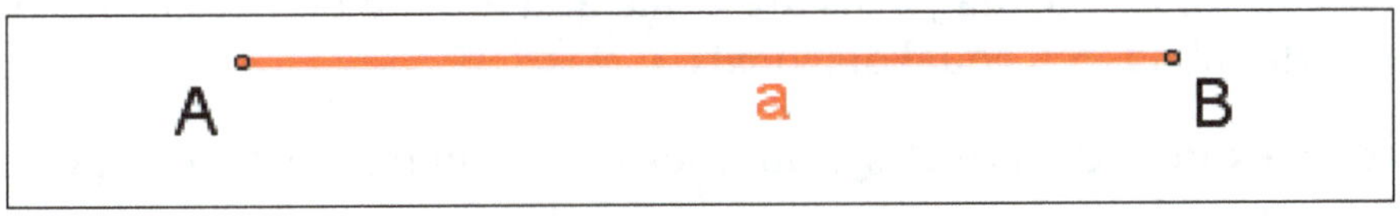

Square at a given side length,
right angle by using Thales' theorem.

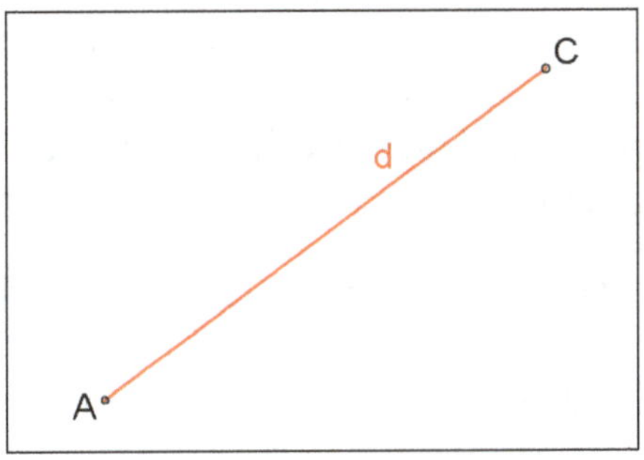

Square at a given diagonal.

Symmetry

The *square* has Dih_4 symmetry, order 8. There are 2 dihedral subgroups: Dih_2, Dih_1, and 3 cyclic subgroups: Z_4, Z_2, and Z_1.

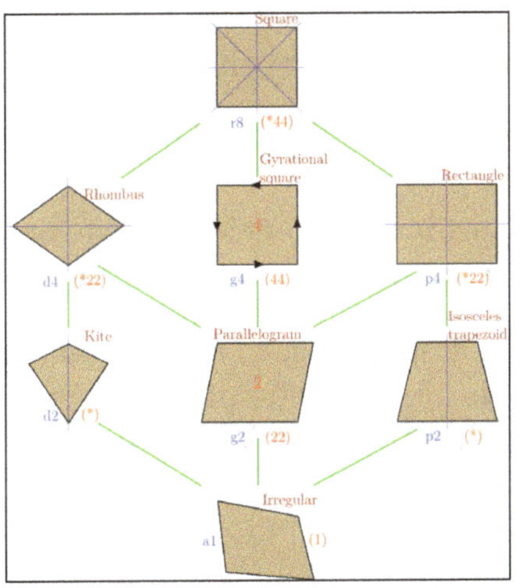

The dihedral symmetries are divided depending on whether they pass through vertices (**d** for diagonal) or edges (**p** for perpendiculars) Cyclic symmetries in the middle column are labeled as **g** for their central gyration orders. Full symmetry of the square is **r12** and no symmetry is labeled **a1**.

A square is a special case of many lower symmetry convex quadrilaterals:

- A rectangle with two adjacent equal sides.
- A quadrilateral with four equal sides and four right angles.
- A parallelogram with one right angle and two adjacent equal sides.
- A rhombus with a right angle.
- A rhombus with all angles equal.
- A quadrilateral where the diagonals are equal and are the perpendicular bisectors of each other, i.e. a rhombus with equal diagonals.

These 6 symmetries express 8 distinct symmetries on a square. John Conway labels these by a letter and group order.

Each subgroup symmetry allows one or more degrees of freedom for irregular quadrilaterals. r8 is full symmetry of the square, and a1 is no symmetry. d4, is the symmetry of a rectangle and p4, is the symmetry of a rhombus. These two forms are duals of each other and have half the symmetry order of the square. d2 is the symmetry of an isosceles trapezoid, and p2 is the symmetry of a kite. g2 defines the geometry of a parallelogram.

Only the g4 subgroup has no degrees of freedom but can seen as a square with directed edges.

Squares Inscribed in Triangles

Every acute triangle has three inscribed squares (squares in its interior such that all four of a square's vertices lie on a side of the triangle, so two of them lie on the same side and hence one side of the square coincides with part of a side of the triangle). In a right triangle two of the squares coincide and have a vertex at the triangle's right angle, so a right triangle has only two *distinct* inscribed squares. An obtuse triangle has only one inscribed square, with a side coinciding with part of the triangle's longest side.

The fraction of the triangle's area that is filled by the square is no more than 1/2.

Squaring the Circle

Squaring the circle is the problem, proposed by ancient geometers, of constructing a square with the same area as a given circle by using only a finite number of steps with compass and straightedge.

In 1882, the task was proven to be impossible, as a consequence of the Lindemann–Weierstrass theorem which proves that pi (π) is a transcendental number, rather than an algebraic irrational number; that is, it is not the root of any polynomial with rational coefficients.

Non-Euclidean Geometry

In non-Euclidean geometry, squares are more generally polygons with 4 equal sides and equal angles.

In spherical geometry, a square is a polygon whose edges are great circle arcs of equal distance, which meet at equal angles. Unlike the square of plane geometry, the angles of such a square are larger than a right angle. Larger spherical squares have larger angles.

In hyperbolic geometry, squares with right angles do not exist. Rather, squares in hyperbolic geometry have angles of less than right angles. Larger hyperbolic squares have smaller angles.

Examples

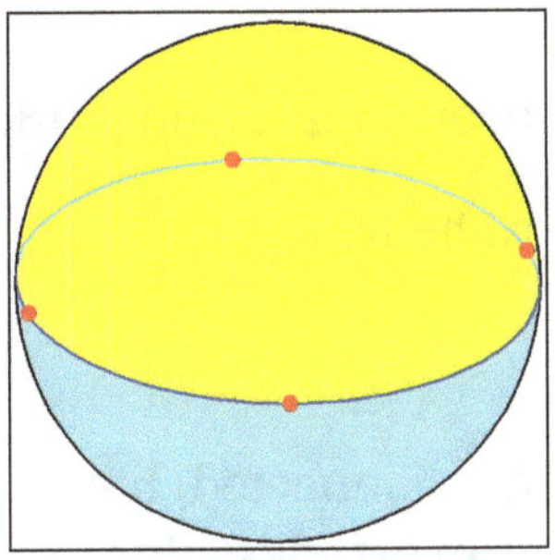

Two squares can tile the sphere with 2 squares around each vertex and 180-degree internal angles. Each square covers an entire hemisphere and their vertices lie along a great circle. This is called a spherical square dihedron. The Schläfli symbol is {4,2}.

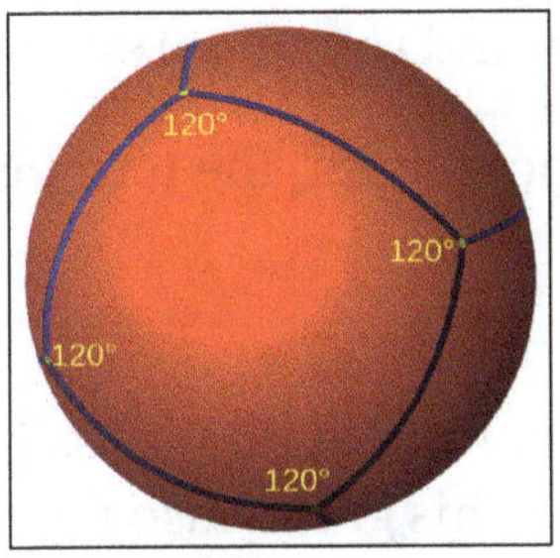

Six squares can tile the sphere with 3 squares around each vertex and 120-degree internal angles. This is called a spherical cube. The Schläfli symbol is {4,3}.

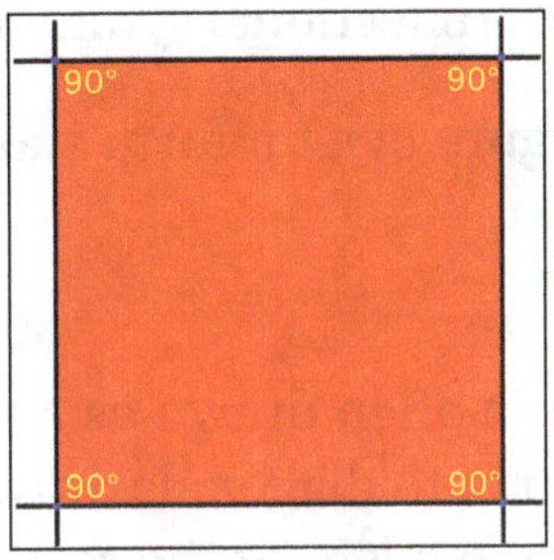

Squares can tile the Euclidean plane with 4 around each vertex, with each square having an internal angle of 90°. The Schläfli symbol is {4,4}.

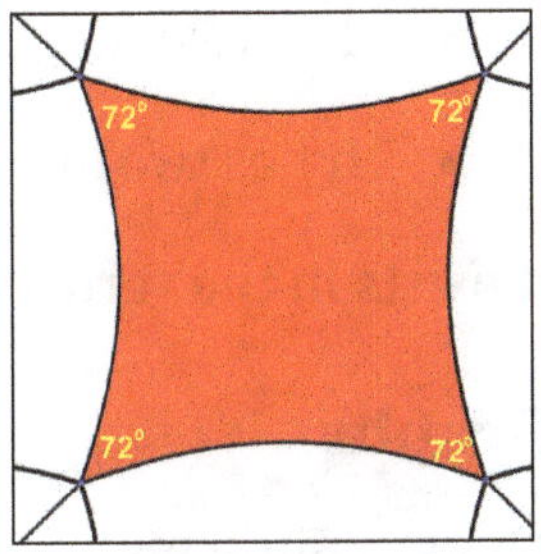

Squares can tile the hyperbolic plane with 5 around each vertex, with each square having 72-degree internal angles. The Schläfli symbol is {4,5}. In fact, for any n ≥ 5 there is a hyperbolic tiling with n squares about each vertex.

Crossed Square

A crossed square is a faceting of the square, a self-intersecting polygon created by removing two opposite edges of a square and reconnecting by its two diagonals. It has half the symmetry of the square, Dih_2, order 4. It has the same vertex arrangement as the square, and is vertex-transitive. It appears as two 45-45-90 triangle with a common vertex, but the geometric intersection is not considered a vertex.

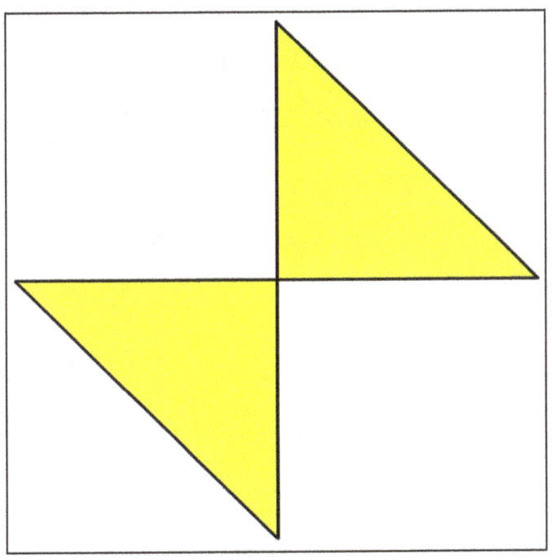

Crossed-square

A crossed square is sometimes likened to a bow tie or butterfly. the crossed rectangle is related, as a faceting of the rectangle, both special cases of crossed quadrilaterals.

The interior of a crossed square can have a polygon density of ±1 in each triangle, dependent upon the winding orientation as clockwise or counterclockwise.

A square and a crossed square have the following properties in common:

- Opposite sides are equal in length.
- The two diagonals are equal in length.
- It has two lines of reflectional symmetry and rotational symmetry of order 2 (through 180°).

It exists in the vertex figure of a uniform star polyhedra, the tetrahemihexahedron.

Graphs

The K_4 complete graph is often drawn as a square with all 6 possible edges connected, hence appearing as a square with both diagonals drawn. This graph also represents an orthographic projection of the 4 vertices and 6 edges of the regular 3-simplex (tetrahedron).

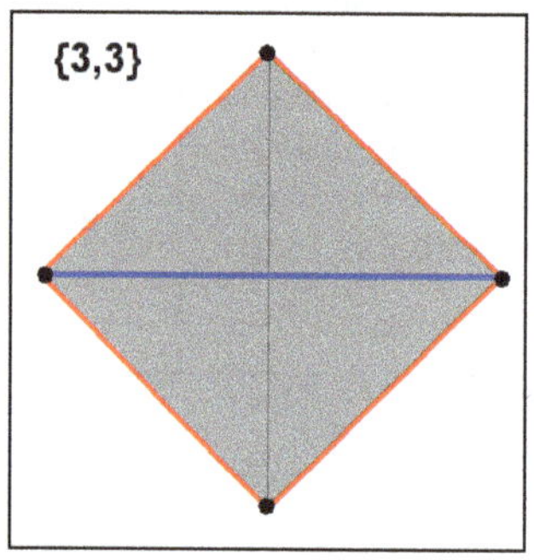

3-simplex (3D)

Trapezoid

Trapezoid (AmE)
Trapezium (BrE)

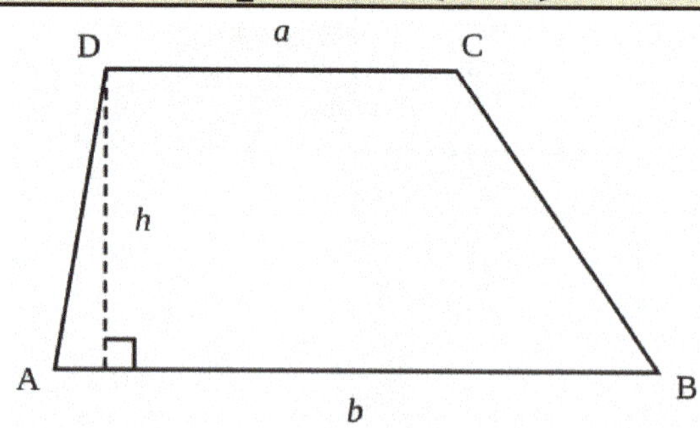

Trapezoid

In Euclidean geometry, a convex quadrilateral with at least one pair of parallel sides is referred to as a trapezoid in American and Canadian English but as a trapezium in English outside North America. The parallel sides are called the *bases* of the trapezoid and the other two sides are called the *legs* or the lateral sides (if they are not parallel; otherwise there are two pairs of bases). A *scalene trapezoid* is a trapezoid with no sides of equal measure, in contrast to the special cases below.

Etymology

The term *trapezium* has been in use in English since 1570, from Late Latin *trapezium*, from Greek τραπέζιον (*trapézion*), literally "a little table", a diminutive of τράπεζα (*trápeza*), "a table", itself from τετράς (*tetrás*), "four" + πέζα (*péza*), "a foot, an edge". The first recorded use of the Greek word translated *trapezoid* (τραπεζοειδή, *trapezoeidé*, "table-like") was by Marinus Proclus (412 to 485 AD) in his Commentary on the first book of Euclid's Elements.

This article uses the term *trapezoid* in the sense that is current in the United States and Canada. In many other languages using a word derived from the Greek for this figure, the form closest to *trapezium* (e.g. French *trapèze*, Italian *trapezio*, Spanish *trapecio*, German *Trapez*, Russian *трапеция*) is used.

Special Cases

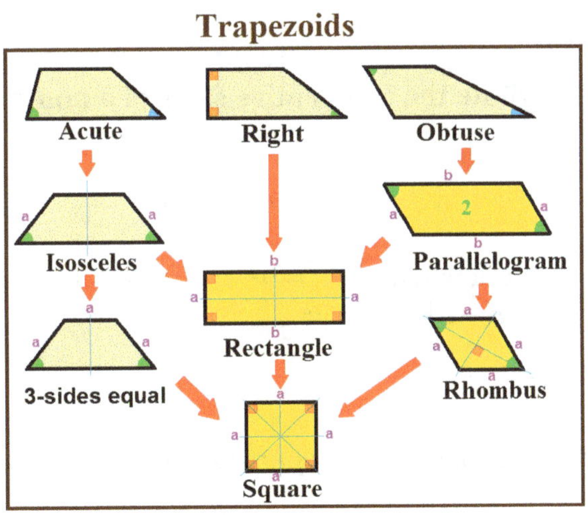

Trapezoid special cases.

A right trapezoid (also called *right-angled trapezoid*) has two adjacent right angles. Right trapezoids are used in the trapezoidal rule for estimating areas under a curve.

An acute trapezoid has two adjacent acute angles on its longer *base* edge, while an obtuse trapezoid has one acute and one obtuse angle on each *base*.

An *acute trapezoid* is also an isosceles trapezoid, if its sides (legs) have the same length, and the base angles have the same measure. It has reflection symmetry.

An *obtuse trapezoid* with two pairs of parallel sides is a parallelogram. A parallelogram has central 2-fold rotational symmetry (or point reflection symmetry).

A Saccheri quadrilateral is similar to a trapezoid in the hyperbolic plane, with two adjacent right angles, while it is a rectangle in the Euclidean plane. A Lambert quadrilateral in the hyperbolic plane has 3 right angles.

A tangential trapezoid is a trapezoid that has an incircle.

Inclusive vs Exclusive Definition

There is some disagreement whether parallelograms, which have two pairs of parallel sides, should be regarded as trapezoids. Some define a trapezoid as a quadrilateral having *only* one pair of parallel sides (the exclusive definition), thereby excluding parallelograms. Others define a trapezoid as a quadrilateral with *at least* one pair of parallel sides (the inclusive definition), making the parallelogram a special type of trapezoid. The latter definition is consistent with its uses in higher mathematics such as calculus. The former definition would make such concepts as the trapezoidal approximation to a definite integral ill-defined. This article uses the inclusive definition and considers parallelograms as special cases of a trapezoid. This is also advocated in the taxonomy of quadrilaterals.

Under the inclusive definition, all parallelograms (including rhombuses, rectangles and squares) are trapezoids. Rectangles have mirror symmetry on mid-edges; rhombuses have mirror symmetry on vertices, while squares have mirror symmetry on both mid-edges and vertices.

Condition of Existence

Four lengths a, c, b, d can constitute the consecutive sides of a non-parallelogram trapezoid with a and b parallel only when,

$$|d-c|<|b-a|<d+c.$$

The quadrilateral is a parallelogram when $d-c=b-a=0$, but it is an ex-tangential quadrilateral (which is not a trapezoid) when $|d-c|=|b-a|\neq 0$.

Characterizations

Given a convex quadrilateral, the following properties are equivalent, and each implies that the quadrilateral is a trapezoid:

- It has two adjacent angles that are supplementary, that is, they add up to 180 degrees.
- The angle between a side and a diagonal is equal to the angle between the opposite side and the same diagonal.
- The diagonals cut each other in mutually the same ratio (this ratio is the same as that between the lengths of the parallel sides).
- The diagonals cut the quadrilateral into four triangles of which one opposite pair are similar.
- The diagonals cut the quadrilateral into four triangles of which one opposite pair have equal areas.

- The product of the areas of the two triangles formed by one diagonal equals the product of the areas of the two triangles formed by the other diagonal.
- The areas S and T of some two opposite triangles of the four triangles formed by the diagonals satisfy the equation,

$$\sqrt{K} = \sqrt{S} + \sqrt{T},$$

where K is the area of the quadrilateral.

- The midpoints of two opposite sides and the intersection of the diagonals are collinear.
- $\sin A \sin C = \sin B \sin D.$
- The cosines of two adjacent angles sum to 0, as do the cosines of the other two angles.
- The cotangents of two adjacent angles sum to 0, as do the cotangents of the other two adjacent angles.
- One bimedian divides the quadrilateral into two quadrilaterals of equal areas.
- Twice the length of the bimedian connecting the midpoints of two opposite sides equals the sum of the lengths of the other sides.

Additionally, the following properties are equivalent, and each implies that opposite sides a and b are parallel:

- The consecutive sides a, c, b, d and the diagonals p, q satisfy the equation,

$$p^2 + q^2 = c^2 + d^2 + 2ab.$$

- The distance v between the midpoints of the diagonals satisfies the equation,

$$v = \frac{|a-b|}{2}.$$

Midsegment and Height

The *midsegment* (also called the median or midline) of a trapezoid is the segment that joins the midpoints of the legs. It is parallel to the bases. Its length m is equal to the average of the lengths of the bases a and b of the trapezoid,

$$m = \frac{a+b}{2}.$$

The midsegment of a trapezoid is one of the two bimedians (the other bimedian divides the trapezoid into equal areas).

The *height* (or altitude) is the perpendicular distance between the bases. In the case that the two bases have different lengths ($a \neq b$), the height of a trapezoid h can be determined by the length of its four sides using the formula,

$$h=\frac{\sqrt{(-a+b+c+d)(a-b+c+d)(a-b+c-d)(a-b-c+d)}}{2|b-a|}$$

where c and d are the lengths of the legs.

Area

The area K of a trapezoid is given by,

$$K=\frac{a+b}{2}\cdot h=mh$$

where a and b are the lengths of the parallel sides, h is the height (the perpendicular distance between these sides), and m is the arithmetic mean of the lengths of the two parallel sides. In 499 AD Aryabhata, a great mathematician-astronomer from the classical age of Indian mathematics and Indian astronomy, used this method in the *Aryabhatiya* (section 2.8). This yields as a special case the well-known formula for the area of a triangle, by considering a triangle as a degenerate trapezoid in which one of the parallel sides has shrunk to a point.

The 7th-century Indian mathematician Bhāskara I derived the following formula for the area of a trapezoid with consecutive sides a, c, b, d:

$$K=\frac{1}{2}(a+b)\sqrt{c^2-\frac{1}{4}\left((b-a)+\frac{c^2-d^2}{b-a}\right)^2}$$

where a and b are parallel and $b > a$. This formula can be factored into a more symmetric version

$$K=\frac{a+b}{4|b-a|}\sqrt{(-a+b+c+d)(a-b+c+d)(a-b+c-d)(a-b-c+d)}.$$

When one of the parallel sides has shrunk to a point (say a = 0), this formula reduces to Heron's formula for the area of a triangle.

Another equivalent formula for the area, which more closely resembles Heron's formula, is

$$K=\frac{a+b}{|b-a|}\sqrt{(s-b)(s-a)(s-b-c)(s-b-d)},$$

where $s=\frac{1}{2}(a+b+c+d)$ is the semiperimeter of the trapezoid. (This formula is similar to Brahmagupta's formula, but it differs from it, in that a trapezoid might not be cyclic (inscribed in a circle). The formula is also a special case of Bretschneider's formula for a general quadrilateral).

From Bretschneider's formula, it follows that,

$$K = \sqrt{\frac{(ab^2 - a^2b - ad^2 + bc^2)(ab^2 - a^2b - ac^2 + bd^2)}{(2(b-a))^2} - (\frac{b^2 + d^2 - a^2 - c^2}{4})^2}.$$

The line that joins the midpoints of the parallel sides, bisects the area.

Diagonals

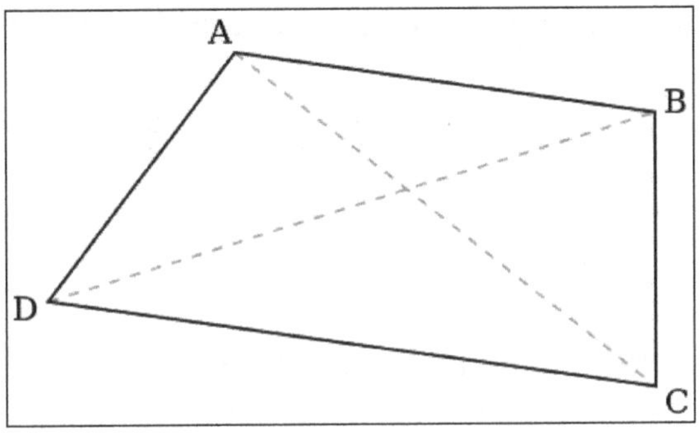

The lengths of the diagonals are,

$$p = \sqrt{\frac{ab^2 - a^2b - ac^2 + bd^2}{b-a}},$$

$$q = \sqrt{\frac{ab^2 - a^2b - ad^2 + bc^2}{b-a}}$$

where a and b are the bases, c and d are the other two sides, and $a < b$.

If the trapezoid is divided into four triangles by its diagonals AC and BD (as shown on the right), intersecting at O, then the area of $\triangle AOD$ is equal to that of $\triangle BOC$, and the product of the areas of $\triangle AOD$ and $\triangle BOC$ is equal to that of $\triangle AOB$ and $\triangle COD$. The ratio of the areas of each pair of adjacent triangles is the same as that between the lengths of the parallel sides.

Let the trapezoid have vertices A, B, C, and D in sequence and have parallel sides AB and DC. Let E be the intersection of the diagonals, and let F be on side DA and G be on side BC such that FEG is parallel to AB and CD. Then FG is the harmonic mean of AB and DC:

$$\frac{1}{FG} = \frac{1}{2}\left(\frac{1}{AB} + \frac{1}{DC}\right).$$

The line that goes through both the intersection point of the extended nonparallel sides and the intersection point of the diagonals, bisects each base.

Other Properties

The center of area (center of mass for a uniform lamina) lies along the line segment joining the midpoints of the parallel sides, at a perpendicular distance x from the longer side b given by,

$$x = \frac{h}{3}\left(\frac{2a+b}{a+b}\right).$$

The center of area divides this segment in the ratio (when taken from the short to the long side),

$$\frac{a+2b}{2a+b}.$$

If the angle bisectors to angles A and B intersect at P, and the angle bisectors to angles C and D intersect at Q, then,

$$PQ = \frac{|AD + BC - AB - CD|}{2}.$$

More on Terminology

The term *trapezoid* was once defined as a quadrilateral without any parallel sides in Britain and elsewhere. (The Oxford English Dictionary says "Often called by English writers in the 19th century".) According to the *Oxford English Dictionary*, the sense of a figure with no sides parallel is the meaning for which Proclus introduced the term "trapezoid". This is retained in the French *trapézoïde* (), German *Trapezoid*, and in other languages. However, this particular sense is considered obsolete.

A *trapezium* in Proclus' sense is a quadrilateral having one pair of its opposite sides parallel. This was the specific sense in England in 17th and 18th centuries, and again the prevalent one in recent use outside North America. A trapezium as any quadrilateral more general than a parallelogram is the sense of the term in Euclid.

Confusingly, the word *trapezium* was sometimes used in England from c. 1800 to c. 1875, to denote an irregular quadrilateral having no sides parallel. This is now obsolete in England, but continues in North America. However this shape is more usually (and less confusingly) just called an irregular quadrilateral.

Application in Geometry

The crossed ladders problem is the problem of finding the distance between the parallel sides of a right trapezoid, given the diagonal lengths and the distance from the perpendicular leg to the diagonal intersection.

Architecture

The Temple of Dendur in the Metropolitan Museum of Art in New York City.

In architecture the word is used to refer to symmetrical doors, windows, and buildings built wider at the base, tapering toward the top, in Egyptian style. If these have straight sides and sharp angular corners, their shapes are usually isosceles trapezoids. This was the standard style for the doors and windows of the Inca.

Application in Biology

In morphology, taxonomy and other descriptive disciplines in which a term for such shapes is necessary, terms such as *trapezoidal* or *trapeziform* commonly are useful in descriptions of particular organs or forms.

Example of a trapeziform pronotum outlined on a spurge bug.

References

- John L. Capinera (11 August 2008). Encyclopedia of Entomology. Springer Science & Business Media. pp. 386, 1062, 1247. ISBN 978-1-4020-6242-1
- Tom M. Apostol and Mamikon A. Mnatsakanian (December 2004). "Figures Circumscribing Circles" (PDF). American Mathematical Monthly: 853–863. Retrieved 26, July 2020
- Owen Byer; Felix Lazebnik; Deirdre L. Smeltzer (19 August 2010). Methods for Euclidean Geometry. MAA. pp. 53–. ISBN 978-0-88385-763-2. Retrieved 07, April 2020
- Park, Poo-Sung. "Regular polytope distances", Forum Geometricorum 16, 2016, 227-232. http://forumgeom.fau.edu/FG2016volume16/FG201627.pdf
- Zalman Usiskin and Jennifer Griffin, "The Classification of Quadrilaterals. A Study of Definition", Information Age Publishing, 2008, p. 59, ISBN 1-59311-695-0

7

Polyhedron, Cone and Sphere

This chapter's focus is on three-dimensional, solid figures in geometry. These figures are all solid, three-dimensional shapes with multiple polygonal faces. In the case of polyhedrons distinctive flat and curved surfaces, pointed at the top as in cones or round solids are defined in three-dimensional space like spheres. This chapter aims to provide a better understanding by closely examining these topics.

Polyhedron

In elementary geometry, a polyhedron (plural polyhedra or polyhedrons) is a solid in three dimensions with flat polygonal faces, straight edges and sharp corners or vertices.

Some Polyhedra

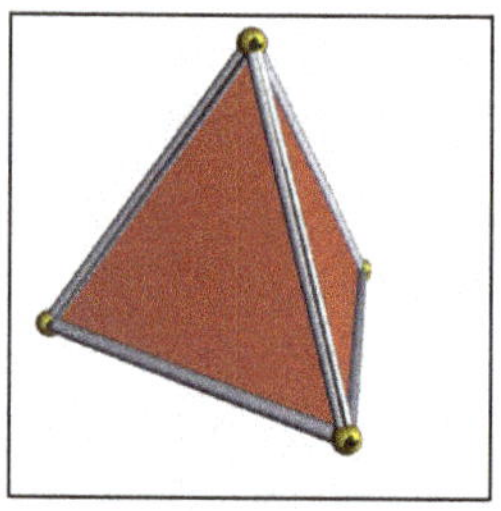

Regular tetrahedron

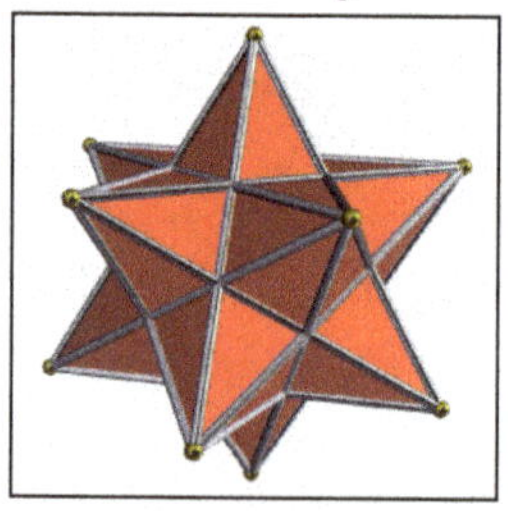

Small stellated dodecahedron

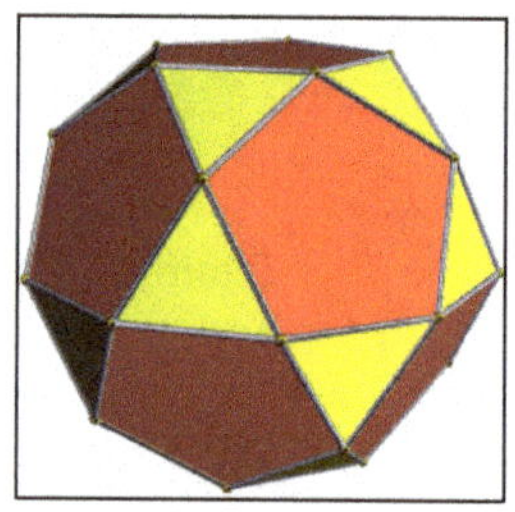

Icosidodecahedron

Great cubicuboctahedron

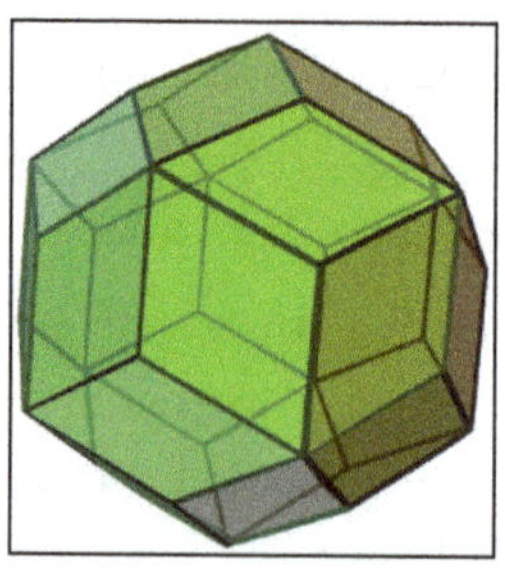

Rhombic triacontahedron

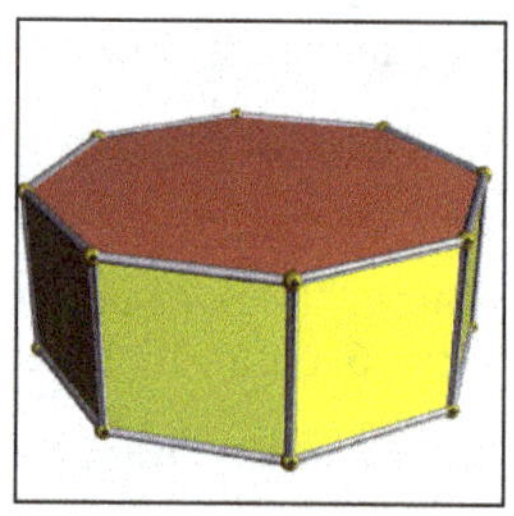

Octagonal prism

Cubes and pyramids are examples of polyhedra.

A polyhedron is said to be convex if its surface (comprising its faces, edges and vertices) does not intersect itself and the line segment joining any two points of the polyhedron is contained in the interior or surface.

A polyhedron is a 3-dimensional example of the more general polytope in any number of dimensions.

Basis for Definition

In elementary geometry, the faces are polygons – regions of planes – meeting in pairs along their edges which are straight-line segments, and with the edges meeting in vertex points. Treating a polyhedron as a solid bounded by flat faces and straight edges is not very precise; for example it is difficult to reconcile with star polyhedra. Grünbaum (1994, p. 43) observed, "The Original Sin in the theory of polyhedra goes back to Euclid, and through Kepler, Poinsot, Cauchy and many others ... [in that] at each stage ... the writers failed to define what are the 'polyhedra' ...." Many definitions of "polyhedron" have been given within particular contexts, some more rigorous than others. For example, definitions based on the idea of a bounding surface rather than a solid are common. However such definitions are not always compatible in other mathematical contexts.

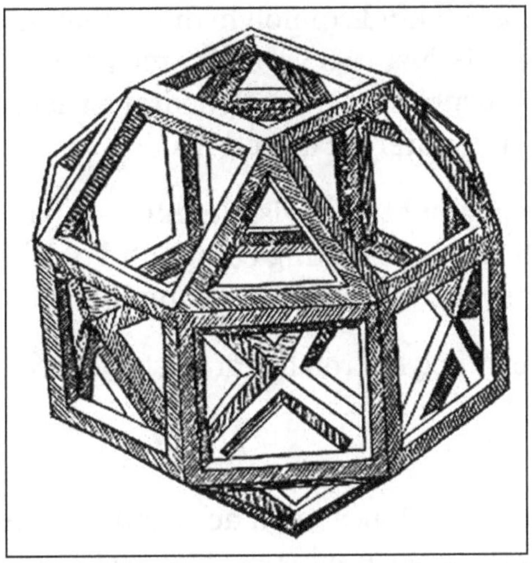

A skeletal polyhedron (specifically, a rhombicuboctahedron) drawn by Leonardo da Vinci to illustrate a book by Luca Pacioli.

One modern approach treats a geometric polyhedron as an injection into real space, a *realisation*, of some abstract polyhedron. Any such polyhedron can be built up from different kinds of element or entity, each associated with a different number of dimensions:

- 3 dimensions: The interior is the volume bounded by the faces. It might or might not be realised as a solid body.
- 2 dimensions: A face is a *polygon* bounded by a circuit of edges, and usually also realises the flat (plane) region inside the boundary. These polygonal faces together make up the polyhedral surface.
- 1 dimension: An edge joins one vertex to another and one face to another, and is usually a line segment. The edges together make up the polyhedral skeleton.
- 0 dimensions: A vertex (plural vertices) is a corner point.

Different approaches - and definitions - may require different realisations. Sometimes the interior volume is considered to be part of the polyhedron, sometimes only the surface is considered, and occasionally only the skeleton of edges or even just the set of vertices.

In such elementary geometric and set-based definitions, a polyhedron is typically understood as a three-dimensional example of the more general polytope in any number of dimensions. For example, a polygon has a two-dimensional body and no faces, while a 4-polytope has a four-dimensional body and an additional set of three-dimensional "cells".

In other mathematical disciplines, the term "polyhedron" may be used to refer to a variety of specialised constructs, some geometric and others purely algebraic or abstract. In such contexts definition of the term "polyhedron" may not be consistent with a polytope but rather in contrast to it.

Characteristics

Polyhedral Surface

A defining characteristic of almost all kinds of polyhedra is that just two faces join along any common edge. Likewise any edge meets just two vertices, one at each end. These two characteristics are dual to each other and they ensure that the polyhedral surface is continuously connected and does not end abruptly or split off in different directions.

For similar reasons, the surface may not be divisible into two parts such that each part is a valid polyhedron. This rules out both self-intersecting compound polyhedra or figures joined only by a vertex or an edge, such as two tetrahedra joined at a common apex.

Every simple (non-self-intersecting) polyhedron has at least two faces with the same number of edges.

Number of Faces

Polyhedra may be classified and are often named according to the number of faces. The naming system is based on Classical Greek, for example tetrahedron (4), pentahedron (5), hexahedron (6), triacontahedron (30), and so on.

For a complete list of the Greek numeral prefixes Numeral prefixes>Table of number prefixes in English>Greek>Quantitative.

Topological Characteristics

The topological class of a polyhedron is defined by its Euler characteristic and orientability.

From this perspective, any polyhedral surface may be classed as certain kind of topological manifold. For example, the surface of a convex or indeed any simply connected polyhedron is a topological sphere.

Euler Characteristic

The Euler characteristic χ relates the number of vertices V, edges E, and faces F of a polyhedron:

$$= V - E + F$$

This is equal to the topological Euler characteristic of its surface. For a convex polyhedron or more generally any simply connected polyhedron (i.e. with surface a topological sphere), $\chi = 2$.

For more complicated shapes, the Euler characteristic relates to the number of toroidal holes, handles and/or cross-caps in the surface and will be less than 2.

Leonhard Euler's discovery of the characteristic which bears his name marked the beginning of the modern discipline of topology.

Orientability

Some polyhedra have two distinct sides to their surface, for example the inside and outside of a convex polyhedron paper model can each be given a different colour (although the inside colour will be hidden from view). We say that the figure is orientable. Some non-convex orientable polyhedra have regions turned "inside out" so that both colours appear on the outside in different places.

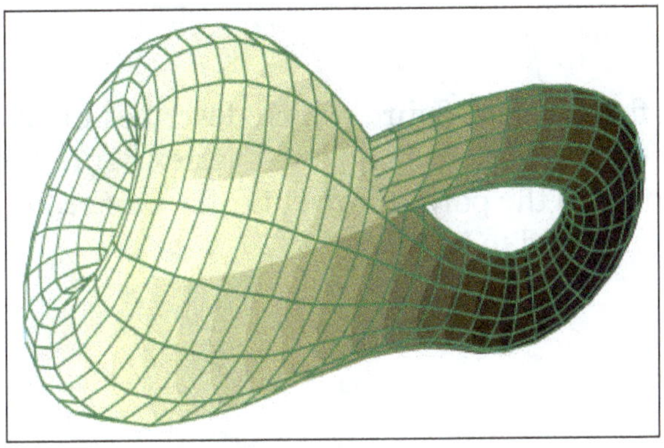

Self-intersecting Klein bottle approximated as quadrilateral polyhedron.

But for some polyhedra, such as the tetrahemihexahedron, this is not possible and the surface is said to be one-sided. Such a polyhedron is non-orientable.

All polyhedra with odd-numbered Euler characteristic χ are non-orientable. A given figure with even $\chi < 2$ may or may not be orientable. For example, the one-holed toroid and the Klein bottle both have $\chi = 0$, with the first being orientable and the other not.

Duality

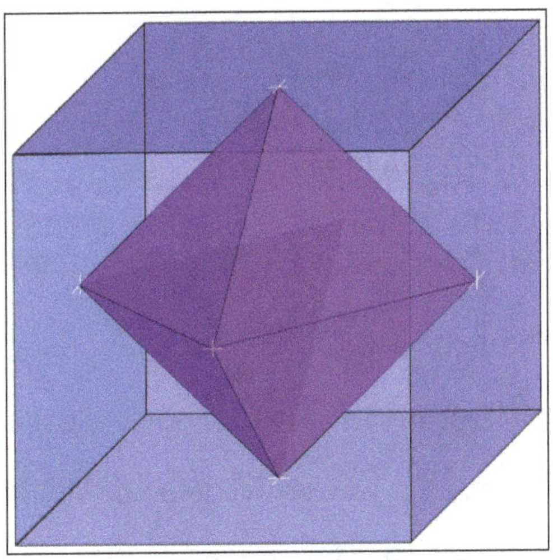

The octahedron is dual to the cube.

For every polyhedron there exists a dual polyhedron having:

- Faces in place of the original's vertices and vice versa,
- The same number of edges,
- The same Euler characteristic and orientability.

The dual of a convex polyhedron and of many other polyhedra can be obtained by the process of polar reciprocation.

Dual polyhedra exist in pairs. The dual of a dual is just the original polyhedron again. Some polyhedra are self-dual, meaning that the dual of the polyhedron is congruent to the original polyhedron.

Vertex Figures

For every vertex one can define a vertex figure, which describes the local structure of the polyhedron around the vertex. Precise definitions vary, but a vertex figure can be thought of as the polygon exposed where a slice through the polyhedron cuts off a corner. If the vertex figure is a regular polygon, then the vertex itself is said to be regular.

Volume

Regular polyhedra:

Any regular polyhedron can be divided up into congruent pyramids, with each pyramid having a face of the polyhedron as its base and the centre of the polyhedron as its apex. The height of a pyramid is equal to the inradius of the polyhedron. If the area of a face is A and the in-radius is r then the volume of the pyramid is one-third of the base times the height, or $Ar/3$. For a regular polyhedron with n faces, its volume is then simply,

$$\text{volume} = nAr/3.$$

For instance, a cube with edges of length L has six faces, each face being a square with area $A = L^2$ The inradius from the center of the face to the center of the cube is $r = L/2$. Then the volume is given by,

$$\text{volume} = \frac{6 \cdot L^2 \cdot \frac{L}{2}}{3} = L^3,$$

the usual formula for the volume of a cube.

Orientable Polyhedra

The volume of any orientable polyhedron can be calculated using the divergence theorem. Consider the vector field $\vec{F}(\vec{x}) = \frac{1}{3}\vec{x} = \left(\frac{x_1}{3}, \frac{x_2}{3}, \frac{x_3}{3}\right)$, whose divergence is identically 1. The divergence theorem implies that the volume is equal to a surface integral of $F(x)$:

$$\text{volume}(\Omega) = \int_{\Omega} \nabla \cdot \vec{F} d\Omega = \oint_{S} \vec{F} \cdot \hat{n} dS.$$

When Ω is the region enclosed by a polyhedron, since the faces of a polyhedron are planar and have piecewise constant normal vectors, this simplifies to,

$$\text{volume} = \frac{1}{3} \sum_{\text{face } i} \vec{x}_i \cdot \hat{n}_i A_i$$

where $\vec{x}_i$ is the ith face's barycenter, $\hat{n}_i$ is its normal vector, and A_i is its area. Once the faces are decomposed in a set of non-overlapping triangles with surface normals pointing away from the volume, the volume is one sixth of the sum over the triple products of the nine Cartesian vertex coordinates of the triangles.

Since it may be difficult to enumerate the faces, volume computation may be challenging, and hence there exist specialized algorithms to determine the volume (many of these generalize to convex polytopes in higher dimensions).

Convex Polyhedra

A polyhedron is said to be convex if its surface (comprising its faces, edges and vertices) does not intersect itself and the line segment joining any two points of the polyhedron is contained in the interior or surface. A convex polyhedron is sometimes defined as a convex set of points in space, the intersection of a set of half-spaces, or the convex hull of a set of points. However many such definitions cannot easily be extended to include self-intersecting figures such as star polyhedra.

Convex polyhedron blocks on display at the Universum museum in Mexico City.

Important classes of convex polyhedra include the highly symmetrical Platonic solids, Archimedean solids and Archimedean duals or Catalan solids, and the regular-faced deltahedra and Johnson solids.

Convex polyhedra, and especially triangular pyramids or 3-simplexes, are important in many areas of mathematics, especially those relating to topology.

Symmetries

Many of the most studied polyhedra are highly symmetrical.

A symmetrical polyhedron can be rotated and superimposed on its original position such that its faces and so on have changed position. All the elements which can be superimposed on each other in this way are said to lie in a given "symmetry orbit". For example, all the faces of a cube lie in one orbit, while all the edges lie in another. If all the elements of a given dimension, say all the faces, lie in the same orbit, the figure is said to be "transitive" on that orbit. For example, a cube has one kind of face so it is face-transitive, while a truncated cube has two kinds of face and is not.

Such polyhedra can be distorted so that they are no longer symmetrical. But where a polyhedral name is given, such as icosidodecahedron, the most symmetrical geometry is almost always implied, unless otherwise stated.

There are several types of highly symmetric polyhedron, classified by which kind of element - faces, edges and/or vertices - belong to a single symmetry orbit:

- Regular if it is vertex-transitive, edge-transitive and face-transitive (this implies that every face is the same regular polygon; it also implies that every vertex is regular).
- Quasi-regular if it is vertex-transitive and edge-transitive (and hence has regular faces) but not face-transitive. A quasi-regular dual is face-transitive and edge-transitive (and hence every vertex is regular) but not vertex-transitive.
- Semi-regular if it is vertex-transitive but not edge-transitive, and every face is a regular polygon. (This is one of several definitions of the term, depending on author. Some definitions overlap with the quasi-regular class). These polyhedra include the semiregular prisms and antiprisms. A semi-regular dual is face-transitive but not vertex-transitive, and every vertex is regular.
- Uniform if it is vertex-transitive and every face is a regular polygon, i.e. it is regular, quasi-regular or semi-regular. A uniform dual is face-transitive and has regular vertices, but is not necessarily vertex-transitive).
- Isogonal or vertex-transitive if all vertices are the same, in the sense that for any two vertices there exists a symmetry of the polyhedron mapping the first isometrically onto the second.
- Isotoxal or edge-transitive if all edges are the same, in the sense that for any two edges there exists a symmetry of the polyhedron mapping the first isometrically onto the second.
- Isohedral or face-transitive if all faces are the same, in the sense that for any two faces there exists a symmetry of the polyhedron mapping the first isometrically onto the second.
- Noble if it is face-transitive and vertex-transitive (but not necessarily edge-transitive). The regular polyhedra are also noble; they are the only noble uniform polyhedra.

A polyhedron can belong to the same overall symmetry group as one of higher symmetry, but will be of lower symmetry if it has several groups of elements in different symmetry orbits. For example, the truncated cube has its triangles and octagons in different orbits.

Some classes of polyhedron have only a single main axis of symmetry. These include the pyramids as well as the semiregular prisms and antiprisms.

Regular Polyhedra

Regular polyhedra are the most highly symmetrical. Altogether there are nine regular polyhedra.

The five convex examples have been known since antiquity and are called the Platonic solids. These are the triangular pyramid or tetrahedron, cube (regular hexahedron), octahedron, dodecahedron and icosahedron:

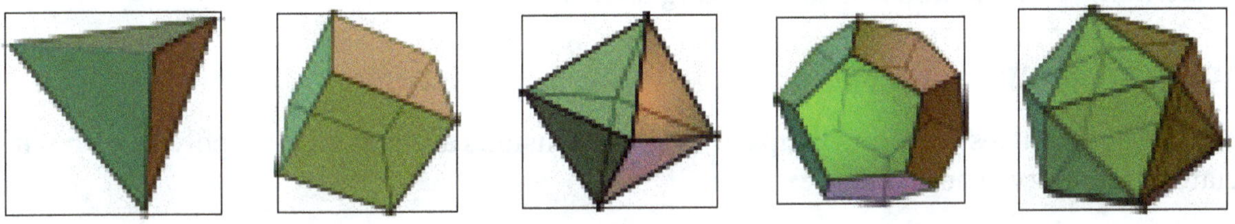

There are also four regular star polyhedra, known as the Kepler-Poinsot polyhedra after their discoverers.

The dual of a regular polyhedron is also regular.

Uniform Polyhedra and their Duals

Uniform polyhedra are *vertex-transitive* and every face is a regular polygon. They may be subdivided into the regular, quasi-regular, or semi-regular, and may be convex or starry.

The uniform duals have irregular faces but are face-transitive and every vertex figure is a regular polygon. A uniform polyhedron has the same symmetry orbits as its dual, with the faces and vertices simply swapped over. The duals of the convex Archimedean polyhedra are sometimes called the Catalan solids.

The uniform polyhedra and their duals are traditionally classified according to their degree of symmetry, and whether they are convex or not.

	Convex uniform	**Convex uniform dual**	**Star uniform**	**Star uniform dual**
Regular	Platonic solids		Kepler–Poinsot polyhedra	
Quasiregular	Archimedean solids	Catalan solids	(no special name)	(no special name)
Semiregular			(no special name)	(no special name)
	Prisms	Bipyramids	Star prisms	Star Bipyramids
	Antiprisms	Trapezohedra	Star antiprisms	Star trapezohedra

Pyramids

Symmetrical pyramids include some of the most time-honoured and famous of all polyhedra, such as the four-sided Egyptian pyramids.

Noble Polyhedra

A noble polyhedron is both isohedral (equal-faced) and isogonal (equal-cornered), but not necessarily equal-sided. Besides the regular polyhedra, there are many other examples.

The dual of a noble polyhedron is also noble.

Isohedra

An isohedron is a polyhedron with symmetries acting transitively on its faces. Their topology can be represented by a face configuration. All 5 Platonic solids and 13 Catalan solids are isohedra, as well as the infinite families of trapezohedra and bipyramids. Some isohedra allow geometric variations including concave and self-intersecting forms.

Symmetry Groups

Many of the symmetries or Point groups in three dimensions are named after polyhedra having the associated symmetry. These include:

- T - chiral tetrahedral symmetry; the rotation group for a regular tetrahedron; order 12.
- T_d - full tetrahedral symmetry; the symmetry group for a regular tetrahedron; order 24.
- T_h - pyritohedral symmetry; the symmetry of a pyritohedron; order 24.
- O - chiral octahedral symmetry;the rotation group of the cube and octahedron; order 24.
- O_h - full octahedral symmetry; the symmetry group of the cube and octahedron; order 48.
- I - chiral icosahedral symmetry; the rotation group of the icosahedron and the dodecahedron; order 60.
- I_h - full icosahedral symmetry; the symmetry group of the icosahedron and the dodecahedron; order 120.
- C_{nv} - *n*-fold pyramidal symmetry.
- D_{nh} - *n*-fold prismatic symmetry.
- D_{nv} - *n*-fold antiprismatic symmetry.

Those with chiral symmetry do not have reflection symmetry and hence have two enantiomorphous forms which are reflections of each other. Examples include the snub cuboctahedron and snub icosidodecahedron.

Polyhedra with Regular Faces

Besides the regular and uniform polyhedra, there are some other classes which have regular faces but lower overall symmetry.

Equal Regular Faces

Convex polyhedra where every face is the same kind of regular polygon may be found among three families:

- Triangles: These polyhedra are called deltahedra. There are eight convex deltahedra, comprising three of the regular (Platonic) polyhedra and five non-uniform examples.

- Squares: The cube is the only convex example. Others can be obtained by joining cubes together, although care must be taken if coplanar faces are to be avoided.
- Pentagons: The regular dodecahedron is the only convex example.

Polyhedra with congruent regular faces of six or more sides are all non-convex, because the vertex of three regular hexagons defines a plane.

The total number of convex polyhedra with equal regular faces is thus ten, comprising the five Platonic solids and the five non-uniform deltahedra.

There are infinitely many non-convex examples. Infinite sponge-like examples called infinite skew polyhedra exist in some of these families.

Johnson Solids

Norman Johnson sought which convex non-uniform polyhedra had regular faces, although not necessarily all alike. In 1966, he published a list of 92 such solids, gave them names and numbers, and conjectured that there were no others. Victor Zalgaller proved in 1969 that the list of these Johnson solids was complete.

Other Important Families of Polyhedra

Stellations and Facettings

Stellation of a polyhedron is the process of extending the faces (within their planes) so that they meet to form a new polyhedron.

It is the exact reciprocal to the process of facetting which is the process of removing parts of a polyhedron without creating any new vertices.

Zonohedra

A zonohedron is a convex polyhedron where every face is a polygon with inversion symmetry or, equivalently, symmetry under rotations through 180°.

Toroidal Polyhedra

A toroidal polyhedron is a polyhedron with an Euler characteristic of 0 or smaller, equivalent to a genus of 1 or greater, representing a torus surface having one or more holes through the middle.

Spacefilling Polyhedra

A spacefilling polyhedron packs with copies of itself to fill space. Such a close-packing or spacefilling is often called a tessellation of space or a honeycomb. Some honeycombs involve more than one kind of polyhedron.

Compounds

A polyhedral compound is made of two or more polyhedra sharing a common centre.

Symmetrical compounds often share the same vertices as other well-known polyhedra and may often also be formed by stellation. Some are listed in the list of Wenninger polyhedron models.

Orthogonal Polyhedra

An orthogonal polyhedron is one all of whose faces meet at right angles, and all of whose edges are parallel to axes of a Cartesian coordinate system. Aside from a rectangular box, orthogonal polyhedra are nonconvex. They are the 3D analogs of 2D orthogonal polygons, also known as rectilinear polygons. Orthogonal polyhedra are used in computational geometry, where their constrained structure has enabled advances on problems unsolved for arbitrary polyhedra, for example, unfolding the surface of a polyhedron to a polygonal net.

Generalisations of Polyhedra

The name 'polyhedron' has come to be used for a variety of objects having similar structural properties to traditional polyhedra.

Apeirohedra

A classical polyhedral surface has a finite number of faces, joined in pairs along edges. If the number of faces extends indefinitely it is called an apeirohedron. Examples include:

- Tilings or tessellations of the plane.
- Sponge-like structures called infinite skew polyhedra.

Complex Polyhedra

A complex polyhedron is one which is constructed in complex Hilbert 3-space. This space has six dimensions: three real ones corresponding to ordinary space, with each accompanied by an imaginary dimension. A complex polyhedron is mathematically more closely related to configurations than to real polyhedra.

Curved Polyhedra

Some fields of study allow polyhedra to have curved faces and edges.

Spherical Polyhedra

The surface of a sphere may be divided by line segments into bounded regions, to form a spherical polyhedron. Much of the theory of symmetrical polyhedra is most conveniently derived in this way.

The first known man-made polyhedra are spherical polyhedra carved in stone, Poinsot used spherical polyhedra to discover the four regular star polyhedra and Coxeter used them to enumerate all but one of the uniform polyhedra.

Some polyhedra, such as hosohedra and dihedra, exist only as spherical polyhedra and have no flat-faced analogue.

Curved Spacefilling Polyhedra

If faces are allowed to be concave as well as convex, adjacent faces may be made to meet together with no gap. Some of these curved polyhedra can pack together to fill space. Two important types are:

- Bubbles in froths and foams, such as Weaire-Phelan bubbles.
- Spacefilling forms used in architecture.

Hollow-faced or Skeletal Polyhedra

It is not necessary to fill in the face of a figure before we can call it a polyhedron. For example, Leonardo da Vinci devised frame models of the regular solids, which he drew for Pacioli's book *Divina Proportione*. In modern times, Branko Grünbaum (1994) made a special study of this class of polyhedra, in which he developed an early idea of abstract polyhedra. He defined a face as a cyclically ordered set of vertices, and allowed faces to be skew as well as planar.

Alternative Usages

From the latter half of the twentieth century, various mathematical constructs have been found to have properties also present in traditional polyhedra. Rather than confining the term "polyhedron" to describe a three-dimensional polytope, it has been adopted to describe various related but distinct kinds of structure.

General Polyhedra

A polyhedron has been defined as a set of points in real affine (or Euclidean) space of any dimension n that has flat sides. It may alternatively be defined as the union of a finite number of convex polyhedra, where a *convex polyhedron* is any point set that is the intersection of a finite number of half-spaces. Unlike an elementary polyhedron, it may be bounded or unbounded. In this meaning, a polytope is a bounded polyhedron.

Analytically, such a convex polyhedron is expressed as the solution set for a system of linear inequalities. Defining polyhedra in this way provides a geometric perspective for problems in linear programming.

Many traditional polyhedral forms are general polyhedra. Other examples include:

- A quadrant in the plane. For instance, the region of the cartesian plane consisting of all points above the horizontal axis and to the right of the vertical axis: $\{ (x, y) : x \geq 0, y \geq 0 \}$. Its sides are the two positive axes, and it is otherwise unbounded.

- An octant in Euclidean 3-space, $\{ (x, y, z) : x \geq 0, y \geq 0, z \geq 0 \}$.
- A prism of infinite extent. For instance a doubly infinite square prism in 3-space, consisting of a square in the *xy*-plane swept along the *z*-axis: $\{ (x, y, z) : 0 \leq x \leq 1, 0 \leq y \leq 1 \}$.
- Each cell in a Voronoi tessellation is a convex polyhedron. In the Voronoi tessellation of a set *S*, the cell *A* corresponding to a point $c \in S$ is bounded (hence a traditional polyhedron) when *c* lies in the interior of the convex hull of *S*, and otherwise (when *c* lies on the boundary of the convex hull of *S*) *A* is unbounded.

Topological Polyhedra

A topological polytope is a topological space given along with a specific decomposition into shapes that are topologically equivalent to convex polytopes and that are attached to each other in a regular way.

Such a figure is called *simplicial* if each of its regions is a simplex, i.e. in an *n*-dimensional space each region has $n+1$ vertices. The dual of a simplicial polytope is called *simple*. Similarly, a widely studied class of polytopes (polyhedra) is that of cubical polyhedra, when the basic building block is an *n*-dimensional cube.

Abstract Polyhedra

An abstract polytope is a partially ordered set (poset) of elements whose partial ordering obeys certain rules of incidence (connectivity) and ranking. The elements of the set correspond to the vertices, edges, faces and so on of the polytope: vertices have rank 0, edges rank 1, etc. with the partially ordered ranking corresponding to the dimensionality of the geometric elements. The empty set, required by set theory, has a rank of −1 and is sometimes said to correspond to the null polytope. An abstract polyhedron is an abstract polytope having the following ranking:

- rank 3: The maximal element, sometimes identified with the body.
- rank 2: The polygonal faces.
- rank 1: The edges.
- rank 0: the vertices.
- rank −1: The empty set, sometimes identified with the *null polytope*.

Any geometric polyhedron is then said to be a “realization” in real space of the abstract poset.

Polyhedra as Graphs

Any polyhedron gives rise to a graph, or skeleton, with corresponding vertices and edges. Thus graph terminology and properties can be applied to polyhedra. For example:

- Due to Steinitz theorem convex polyhedra are in one-to-one correspondence with 3-connected planar graphs.
- The tetrahedron gives rise to a complete graph (K_4). It is the only polyhedron to do so.

- The octahedron gives rise to a strongly regular graph, because adjacent vertices always have two common neighbors, and non-adjacent vertices have four.
- The Archimedean solids give rise to regular graphs: 7 of the Archimedean solids are of degree 3, 4 of degree 4, and the remaining 2 are chiral pairs of degree 5.

History

Prehistory

Polyhedra appeared in early architectural forms such as cubes and cuboids, with the earliest four-sided pyramids of ancient Egypt also dating from the Stone Age.

The Etruscans preceded the Greeks in their awareness of at least some of the regular polyhedra, as evidenced by the discovery near Padua (in Northern Italy) in the late 19th century of a dodecahedron made of soapstone, and dating back more than 2,500 years (Lindemann, 1987).

Greek Civilisation

The earliest known *written* records of these shapes come from Classical Greek authors, who also gave the first known mathematical description of them. The earlier Greeks were interested primarily in the convex regular polyhedra, which came to be known as the Platonic solids. Pythagoras knew at least three of them, and Theaetetus (circa 417 B. C.) described all five. Eventually, Euclid described their construction in his *Elements*. Later, Archimedes expanded his study to the convex uniform polyhedra which now bear his name. His original work is lost and his solids come down to us through Pappus.

China

Cubical gaming dice in China have been dated back as early as 600 B.C.

By 236 AD, Liu Hui was describing the dissection of the cube into its characteristic tetrahedron (orthoscheme) and related solids, using assemblages of these solids as the basis for calculating volumes of earth to be moved during engineering excavations.

Islamic Civilisation

After the end of the Classical era, scholars in the Islamic civilisation continued to take the Greek knowledge forward.

The 9th century scholar Thabit ibn Qurra gave formulae for calculating the volumes of polyhedra such as truncated pyramids.

Then in the 10th century Abu'l Wafa described the convex regular and quasiregular spherical polyhedra.

Renaissance

As with other areas of Greek thought maintained and enhanced by Islamic scholars, Western interest in polyhedra revived during the Italian Renaissance. Artists constructed skeletal polyhedra, depicting them from life as a part of their investigations into perspective. Several appear in marquetry panels of the period. Piero della Francesca gave the first written description of direct geometrical construction of such perspective views of polyhedra. Leonardo da Vinci made skeletal models of several polyhedra and drew illustrations of them for a book by Pacioli. A painting by an anonymous artist of Pacioli and a pupil depicts a glass rhombicuboctahedron half-filled with water.

As the Renaissance spread beyond Italy, later artists such as Wenzel Jamnitzer, Dürer and others also depicted polyhedra of various kinds, many of them novel, in imaginative etchings.

Star Polyhedra

For almost 2,000 years, the concept of a polyhedron as a convex solid had remained as developed by the ancient Greek mathematicians.

During the Renaissance star forms were discovered. A marble tarsia in the floor of St. Mark's Basilica, Venice, depicts a stellated dodecahedron. Artists such as Wenzel Jamnitzer delighted in depicting novel star-like forms of increasing complexity.

Johannes Kepler (1571 - 1630) used star polygons, typically pentagrams, to build star polyhedra. Some of these figures may have been discovered before Kepler's time, but he was the first to recognize that they could be considered "regular" if one removed the restriction that regular polytopes must be convex. Later, Louis Poinsot realised that star vertex figures (circuits around each corner) can also be used, and discovered the remaining two regular star polyhedra. Cauchy proved Poinsot's list complete, and Cayley gave them their accepted English names: (Kepler's) the small stellated dodecahedron and great stellated dodecahedron, and (Poinsot's) the great icosahedron and great dodecahedron. Collectively they are called the Kepler-Poinsot polyhedra.

The Kepler-Poinsot polyhedra may be constructed from the Platonic solids by a process called stellation. Most stellations are not regular. The study of stellations of the Platonic solids was given a big push by H. S. M. Coxeter and others in 1938, with the now famous paper *The 59 icosahedra.*

The reciprocal process to stellation is called facetting (or faceting). Every stellation of one polytope is dual, or reciprocal, to some facetting of the dual polytope. The regular star polyhedra can also be obtained by facetting the Platonic solids. Bridge 1974 listed the simpler facettings of the dodecahedron, and reciprocated them to discover a stellation of the icosahedron that was missing from the set of "59". More have been discovered since, and the story is not yet ended.

Euler's Formula and Topology

Two other modern mathematical developments had a profound effect on polyhedron theory.

In 1750 the German Leonhard Euler for the first time considered the edges of a polyhedron, allowing him to discover his polyhedron formula relating the number of vertices, edges and faces. This signalled the birth of topology, sometimes referred to as "rubber sheet geometry", and the Frenchman Henri Poincaré developed its core ideas around the end of the nineteenth century. This allowed many longstanding issues over what was or was not a polyhedron to be resolved.

Max Brückner summarised work on polyhedra to date, including many findings of his own, in his book "Vielecke und Vielflache: Theorie und Geschichte" (Polygons and polyhedra: Theory and History). Published in German in 1900, it remained little known.

Meanwhile, the discovery of higher dimensions led to the idea of a polyhedron as a three-dimensional example of the more general polytope.

The Twentieth Century Revival

By the early years of the twentieth century, mathematicians had moved on and geometry was little studied. Coxeter's analysis in *The Fifty-Nine Icosahedra* introduced modern ideas from graph theory and combinatorics into the study of polyhedra, signalling a rebirth of interest in geometry.

Coxeter himself went on to enumerate the star uniform polyhedra for the first time, to treat tilings of the plane as polyhedra, to discover the regular skew polyhedra and to develop the theory of complex polyhedra first discovered by Shephard in 1952, as well as making fundamental contributions to many other areas of geometry.

In the second part of the twentieth century, Grünbaum published important works in two areas. One was in convex polytopes, where he noted a tendency among mathematicians to define a "polyhedron" in different and sometimes incompatible ways to suit the needs of the moment. The other was a series of papers broadening the accepted definition of a polyhedron, for example discovering many new Regular polyhedra. At the close of the 20th century these latter ideas merged with other work on incidence complexes to create the modern idea of an abstract polyhedron (as an abstract 3-polytope), notably presented by McMullen and Schulte.

Polyhedra in Nature

Irregular polyhedra appear in nature as crystals.

Cone

A cone is a three-dimensional geometric shape that tapers smoothly from a flat base (frequently, though not necessarily, circular) to a point called the apex or vertex.

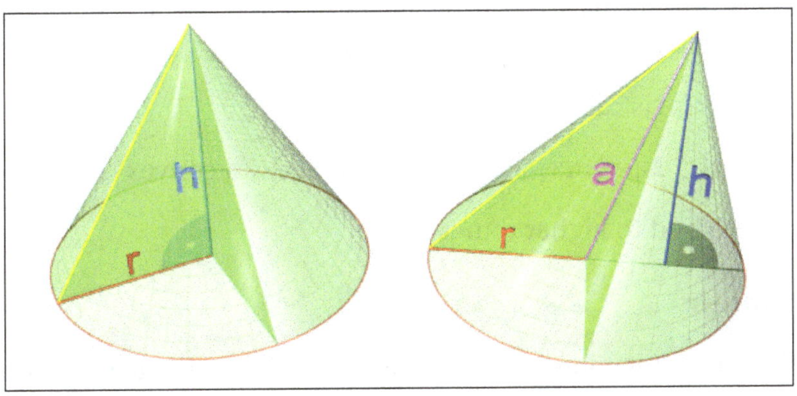

A right circular cone and an oblique circular cone.

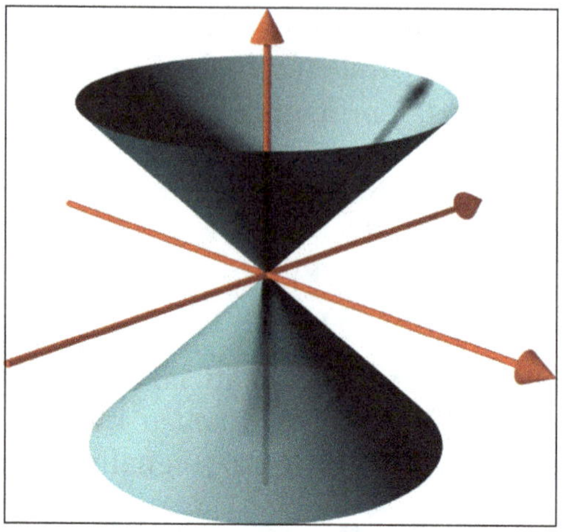

A double cone (not shown infinitely extended).

A cone is formed by a set of line segments, half-lines, or lines connecting a common point, the apex, to all of the points on a base that is in a plane that does not contain the apex. Depending on the author, the base may be restricted to be a circle, any one-dimensional quadratic form in the plane, any closed one-dimensional figure, or any of the above plus all the enclosed points. If the enclosed points are included in the base, the cone is a solid object; otherwise it is a two-dimensional object in three-dimensional space. In the case of a solid object, the boundary formed by these lines or partial lines is called the *lateral surface*; if the lateral surface is unbounded, it is a conical surface.

In the case of line segments, the cone does not extend beyond the base, while in the case of half-lines, it extends infinitely far. In the case of lines, the cone extends infinitely far in both directions from the apex, in which case it is sometimes called a *double cone*. Either half of a double cone on one side of the apex is called a *nappe*.

The axis of a cone is the straight line (if any), passing through the apex, about which the base (and the whole cone) has a circular symmetry.

In common usage in elementary geometry, cones are assumed to be right circular, where *circular* means that the base is a circle and *right* means that the axis passes through the centre of the base at right angles to its plane. If the base is right circular the intersection of a plane with this surface is a conic section. In general, however, the base may be any shape and the apex may lie anywhere (though it is usually assumed that the base is bounded and therefore has finite area, and that the apex lies outside the plane of the base). Contrasted with right cones are oblique cones, in which the axis passes through the centre of the base non-perpendicularly.

A cone with a polygonal base is called a pyramid.

Depending on the context, "cone" may also mean specifically a convex cone or a projective cone.

Cones can also be generalized to higher dimensions.

Further Terminology

The perimeter of the base of a cone is called the "directrix", and each of the line segments between the directrix and apex is a "generatrix" or "generating line" of the lateral surface. (For the connection between this sense of the term "directrix" and the directrix of a conic section, Dandelin spheres).

The "base radius" of a circular cone is the radius of its base; often this is simply called the radius of the cone. The aperture of a right circular cone is the maximum angle between two generatrix lines; if the generatrix makes an angle θ to the axis, the aperture is 2θ.

A cone with a region including its apex cut off by a plane is called a "truncated cone"; if the truncation plane is parallel to the cone's base, it is called a frustum. An "elliptical cone" is a cone with an elliptical base. A "generalized cone" is the surface created by the set of lines passing through a vertex and every point on a boundary.

Measurements and Equations

Volume

The volume V of any conic solid is one third of the product of the area of the base A_B and the height h,

$$V = \frac{1}{3} A_B h.$$

In modern mathematics, this formula can easily be computed using calculus – it is, up to scaling, the integral $\int x^2 dx = \tfrac{1}{3}x^3$. Without using calculus, the formula can be proven by comparing the cone to a pyramid and applying Cavalieri's principle – specifically, comparing the cone to a (vertically scaled) right square pyramid, which forms one third of a cube. This formula cannot be proven without using such infinitesimal arguments – unlike the 2-dimensional formulae for polyhedral area, though similar to the area of the circle – and hence admitted less rigorous proofs before the advent of calculus, with the ancient Greeks using the method of exhaustion. This is essentially the content of Hilbert's third problem – more precisely, not all polyhedral pyramids are *scissors congruent* (can be cut apart into finite pieces and rearranged into the other), and thus volume cannot be computed purely by using a decomposition argument.

Center of Mass

The center of mass of a conic solid of uniform density lies one-quarter of the way from the center of the base to the vertex, on the straight line joining the two.

Right Circular Cone

Volume

For a circular cone with radius R and height H, the formula for volume becomes,

$$V = \int_0^H r^2 \pi \, dh$$

where r is the radius of the cone at height h measured from the apex,

$$r = R\frac{h}{H}.$$

Thus,

$$V = \int_0^H \left[R\frac{h}{H} \right]^2 \pi\, dh$$

and so,

$$V = \frac{1}{3}\pi R^2 H.$$

Surface Area

The lateral surface area of a right circular cone is $LSA = \pi r l$ where r is the radius of the circle at the bottom of the cone and l is the lateral height (the length of a line segment from the apex of the cone along its side to its base) of the cone (given by the Pythagorean theorem $l = \sqrt{r^2 + h^2}$ where h is the height of the cone). The surface area of the bottom circle of a cone is the same as for any circle, πr^2. Thus, the total surface area of a right circular cone can be expressed as each of the following:

- Radius and height,

$$\pi r^2 + \pi r\sqrt{r^2 + h^2}$$

(the area of the base plus the area of the lateral surface; the term $\sqrt{r^2 + h^2}$ is the slant height).

$$\pi r\left(r + \sqrt{r^2 + h^2}\right)$$

where r is the radius and h is the height.

- Radius and lateral height,

$$\pi r^2 + \pi r l$$

$$\pi r(r + l)$$

where r is the radius and l is the lateral height.

- Circumference and lateral height,

$$\frac{c^2}{4\pi} + \frac{cl}{2}$$

$$\left(\frac{c}{2}\right)\left(\frac{c}{2\pi}+l\right)$$

where c is the circumference and l is the lateral height.

- Apex angle and height,

$$\pi h^2 \tan\frac{\text{È}}{2}(\tan\frac{\text{È}}{2}+\sec\frac{\text{È}}{2})$$

where Θ is the apex angle and h is the height.

Equation Form

A right solid circular cone with height h and aperture 2θ, whose axis is the z coordinate axis and whose apex is the origin, is described parametrically as,

$$F(s,t,u)=\left(u\tan s\cos t, u\tan s\sin t, u\right)$$

where s,t,u range over $[0,\theta)$, $[0,2\pi)$, and $[0,h]$, respectively.

In implicit form, the same solid is defined by the inequalities,

$$\{F(x,y,z)\leq 0, z\geq 0, z\leq h\},$$

where

$$F(x,y,z)=(x^2+y^2)(\cos\theta)^2-z^2(\sin\theta)^2.$$

More generally, a right circular cone with vertex at the origin, axis parallel to the vector d, and aperture 2θ, is given by the implicit vector equation $F(u)=0$ where,

$$F(u)=(u\cdot d)^2-(d\cdot d)(u\cdot u)(\cos\theta)^2 \quad \text{or} \quad F(u)=u\cdot d-|d||u|\cos\theta$$

where $u=(x,y,z)$, and $u\cdot d$ denotes the dot product.

Elliptic Cone

In the Cartesian coordinate system, an *elliptic cone* is the locus of an equation of the form,

$$\frac{x^2}{a^2}+\frac{y^2}{b^2}=\frac{z^2}{c^2}.$$

Projective Geometry

In projective geometry, a cylinder is simply a cone whose apex is at infinity. Intuitively, if one keeps the base fixed and takes the limit as the apex goes to infinity, one obtains a cylinder, the angle of the side increasing as arctan, in the limit forming a right angle. This is useful in the definition of degenerate conics, which require considering the cylindrical conics.

In projective geometry, a cylinder is simply a cone whose apex is at infinity, which corresponds visually to a cylinder in perspective appearing to be a cone towards the sky.

Higher Dimensions

The definition of a cone may be extended to higher dimensions. In this case, one says that a convex set C in the real vector space R^n is a cone (with apex at the origin) if for every vector x in C and every nonnegative real number a, the vector ax is in C. In this context, the analogues of circular cones are not usually special; in fact one is often interested in *polyhedral cones*.

Sphere

A sphere is a perfectly round geometrical object in three-dimensional space that is the surface of a completely round ball, (viz., analogous to a circular object in two dimensions).

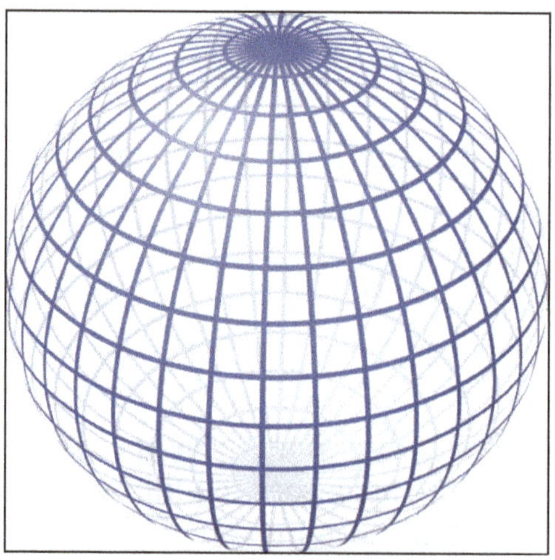

A two-dimensional perspective projection of a sphere.

r – radius of the sphere.

Like a circle, which geometrically is a two-dimensional object, a sphere is defined mathematically as the set of points that are all at the same distance r from a given point, but in three-dimensional space. This distance r is the radius of the ball, and the given point is the center of the mathematical ball. The longest straight line through the ball, connecting two points of the sphere, passes through the center and its length is thus twice the radius; it is a diameter of the ball.

While outside mathematics the terms "sphere" and "ball" are sometimes used interchangeably, in mathematics a distinction is made between the sphere (a two-dimensional closed surface embedded in three-dimensional Euclidean space) and the ball (a three-dimensional shape that includes the sphere as well as everything inside the sphere). The ball and the sphere share the same radius, diameter, and center.

Surface Area

The surface area of a sphere is:

$$A = 4\pi r^2.$$

Archimedes first derived this formula from the fact that the projection to the lateral surface of a circumscribed cylinder (for example, the Lambert cylindrical equal-area projection) is area-preserving; it equals the derivative of the formula for the volume with respect to r because the total volume inside a sphere of radius r can be thought of as the summation of the surface area of an infinite number of spherical shells of infinitesimal thickness concentrically stacked inside one another from radius 0 to radius r. At infinitesimal thickness the discrepancy between the inner and outer surface area of any given shell is infinitesimal, and the elemental volume at radius r is simply the product of the surface area at radius r and the infinitesimal thickness.

At any given radius r, the incremental volume (δV) equals the product of the surface area at radius r ($A(r)$) and the thickness of a shell (δr):

$$\delta V \approx A(r)\cdot\delta r.$$

The total volume is the summation of all shell volumes:

$$V \approx \sum A(r)\cdot\ r.$$

In the limit as δr approaches zero this equation becomes:

$$V = \int_0^r A(r')dr'.$$

Substitute *V*:

$$\frac{4}{3}\pi r^3 = \int_0^r A(r')dr'.$$

Differentiating both sides of this equation with respect to *r* yields *A* as a function of *r*:

$$4\pi r^2 = A(r).$$

Which is generally abbreviated as:

$$A = 4\pi r^2.$$

Alternatively, the area element on the sphere is given in spherical coordinates by $dA = r^2 \sin\theta \, d\theta \, d\varphi$. In Cartesian coordinates, the area element is,

$$dS = \frac{r}{\sqrt{r^2 - \sum_{i \neq k} x_i^2}} \prod_{i \neq k} dx_i, \ \forall k.$$

The total area can thus be obtained by integration:

$$A = \int_0^{2\pi} \int_0^{\pi} r^2 \sin\theta d\theta d\varphi = 4\pi r^2.$$

Enclosed Volume

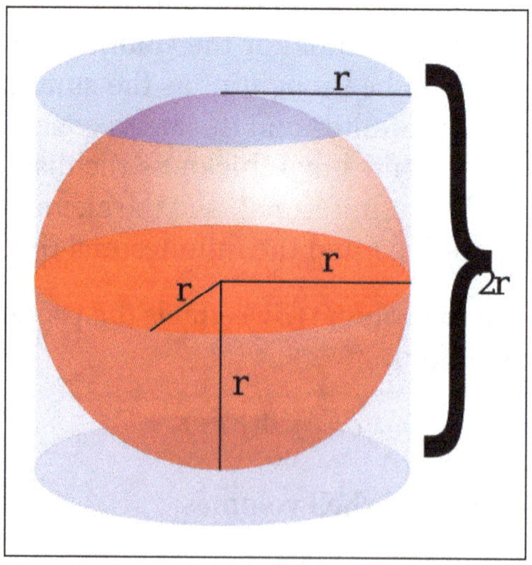

Circumscribed cylinder to a sphere.

In three dimensions, the volume inside a sphere (that is the volume of a ball) is derived to be,

$$V = \frac{4}{3}\pi r^3$$

where r is the radius of the sphere. Archimedes first derived this formula, which shows that the volume inside a sphere is 2/3 that of a circumscribed cylinder. (This assertion follows from Cavalieri's principle.) In modern mathematics, this formula can be derived using integral calculus, i.e. disk integration to sum the volumes of an infinite number of circular disks of infinitesimally small thickness stacked centered side by side along the x-axis from $x = 0$ where the disk has radius r (i.e. $y = r$) to $x = r$ where the disk has radius 0 (i.e. $y = 0$).

At any given x, the incremental volume (δV) equals the product of the cross-sectional area of the disk at x and its thickness (δx):

$$\delta V \approx \pi y^2 \cdot \delta x.$$

The total volume is the summation of all incremental volumes:

$$V \approx \sum \pi y^2 \cdot \delta x.$$

In the limit as δx approaches zero this equation becomes:

$$V = \int_{-r}^{r} \pi y^2 dx.$$

At any given x, a right-angled triangle connects x, y and r to the origin; hence, applying the Pythagorean theorem yields:

$$y^2 = r^2 - x^2.$$

Thus, substituting y with a function of x gives:

$$V = \int_{-r}^{r} \pi (r^2 - x^2) dx.$$

Which can now be evaluated as follows:

$$V = \pi \left[r^2 x - \frac{x^3}{3} \right]_{-r}^{r} = \pi \left(r^3 - \frac{r^3}{3} \right) - \pi \left(-r^3 + \frac{r^3}{3} \right) = \frac{4}{3}\pi r^3.$$

Therefore the volume of a sphere is:

$$V = \frac{4}{3}\pi r^3.$$

Alternatively this formula is found using spherical coordinates, with volume element,

$$dV = r^2 \sin\theta dr d\theta d\varphi$$

so,

$$V = \int_0^{2\pi}\int_0^{\pi}\int_0^{r} r'^2 \sin\theta dr' d\theta d\varphi = \frac{4}{3}\pi r^3$$

For most practical purposes, the volume inside a sphere inscribed in a cube can be approximated as 52.4% of the volume of the cube, since $\pi/6 \approx 0.5236$. For example, a sphere with diameter 1 meter has 52.4% the volume of a cube with edge length 1 meter, or about 0.524 m^3.

In higher dimensions, the analog of a sphere is called a hypersphere, which encloses an n-ball. General recursive and non-recursive formulas exist for the volume of an n-ball.

Equations in Three-dimensional Space

In analytic geometry, a sphere with center (x_0, y_0, z_0) and radius r is the locus of all points (x, y, z) such that,

$$(x - x_0)^2 + (y - y_0)^2 + (z - z_0)^2 = r^2.$$

The points on the sphere with radius r can be parameterized via,

$$\begin{aligned} x &= x_0 + r\cos\theta\sin\varphi \\ y &= y_0 + r\sin\theta\sin\varphi \qquad (0 \le \theta \le 2\pi \text{ and } 0 \le \varphi \le \pi) \\ z &= z_0 + r\cos\varphi \end{aligned}$$

A sphere of any radius centered at zero is an integral surface of the following differential form:

$$xdx + ydy + zdz = 0.$$

This equation reflects that position and velocity vectors of a point traveling on the sphere are always orthogonal to each other.

The sphere has the smallest surface area of all surfaces that enclose a given volume, and it encloses the largest volume among all closed surfaces with a given surface area. The sphere therefore appears in nature: for example, bubbles and small water drops are roughly spherical because the surface tension locally minimizes surface area.

The surface area relative to the mass of a sphere is called the specific surface area and can be expressed from the above stated equations as,

$$\mathrm{SSA} = \frac{A}{V\rho} = \frac{3}{r\rho},$$

where ρ is the density (the ratio of mass to volume).

A sphere can also be defined as the surface formed by rotating a circle about any diameter. Replacing the circle with an ellipse rotated about its major axis, the shape becomes a prolate spheroid; rotated about the minor axis, an oblate spheroid.

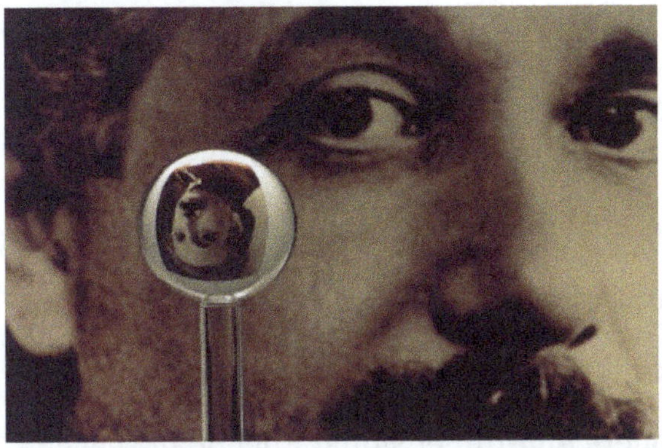

An image of one of the most accurate human-made spheres, as it refracts the image of Einstein in the background. This sphere was a fused quartz gyroscope for the Gravity Probe B experiment, and differs in shape from a perfect sphere by no more than 40 atoms (less than 10 nanometers) of thickness. It was announced on 1 July 2008 that Australian scientists had created even more nearly perfect spheres, accurate to 0.3 nanometers, as part of an international hunt to find a new global standard kilogram.

Terminology

Pairs of points on a sphere that lie on a straight line through the sphere's center are called antipodal points. A great circle is a circle on the sphere that has the same center and radius as the sphere and consequently divides it into two equal parts. The shortest distance along the surface between two distinct non-antipodal points on the surface is on the unique great circle that includes the two points. Equipped with the great-circle distance, a great circle becomes the Riemannian circle.

If a particular point on a sphere is (arbitrarily) designated as its *north pole*, then the corresponding antipodal point is called the *south pole*, and the equator is the great circle that is equidistant to them. Great circles through the two poles are called lines (or meridians) of longitude, and the line connecting the two poles is called the axis of rotation. Circles on the sphere that are parallel to the equator are lines of latitude. This terminology is also used for such approximately spheroidal astronomical bodies as the planet Earth.

Hemisphere

Any plane that includes the center of a sphere divides it into two equal hemispheres. Any two intersecting planes that include the center of a sphere subdivide the sphere into four lunes or biangles, the vertices of which all coincide with the antipodal points lying on the line of intersection of the planes.

The antipodal quotient of the sphere is the surface called the real projective plane, which can also be thought of as the northern hemisphere with antipodal points of the equator identified.

The round hemisphere is conjectured to be the optimal (least area) filling of the Riemannian circle.

The circles of intersection of any plane not intersecting the sphere's center and the sphere's surface are called *spheric sections*.

Generalizations

Dimensionality

Spheres can be generalized to spaces of any number of dimensions. For any natural number n, an "n-sphere," often written as S^n, is the set of points in $(n + 1)$-dimensional Euclidean space that are at a fixed distance r from a central point of that space, where r is, as before, a positive real number. In particular:

- S^0: a 0-sphere is a pair of endpoints of an interval $[-r, r]$ of the real line.
- S^1: a 1-sphere is a circle of radius r.
- S^2: a 2-sphere is an ordinary sphere.
- S^3: a 3-sphere is a sphere in 4-dimensional Euclidean space.

Spheres for $n > 2$ are sometimes called hyperspheres.

The n-sphere of unit radius centered at the origin is denoted S^n and is often referred to as "the" n-sphere. Note that the ordinary sphere is a 2-sphere, because it is a 2-dimensional surface (which is embedded in 3-dimensional space).

The surface area of the $(n - 1)$-sphere of radius 1 is,

$$\frac{2\pi^{n/2}}{\Gamma\left(\frac{n}{2}\right)}$$

where $\Gamma(z)$ is Euler's gamma function.

Another expression for the surface area is,

$$\begin{cases} \dfrac{(2\pi)^{n/2} r^{n-1}}{2\cdot 4\cdots(n-2)}, & \text{if } n \text{ is even;} \\[2ex] \dfrac{2(2\pi)^{(n-1)/2} r^{n-1}}{1\cdot 3\cdots(n-2)}, & \text{if } n \text{ is odd.} \end{cases}$$

and the volume is the surface area times r/n or,

$$\begin{cases} \dfrac{(2\ )^{n/2}\ \ ^{n}}{2\cdot 4}, & \text{if } \ \text{ is even;} \\[2ex] \dfrac{2(2\ )^{(n\ \ 1)/2}\ \ ^{n}}{1\cdot 3}, & \text{if } \ \text{ is odd.} \end{cases}$$

Metric Spaces

More generally, in a metric space (E,d), the sphere of center x and radius $r > 0$ is the set of points y such that $d(x,y) = r$.

If the center is a distinguished point that is considered to be the origin of E, as in a normed space, it is not mentioned in the definition and notation. The same applies for the radius if it is taken to equal one, as in the case of a unit sphere.

Unlike a ball, even a large sphere may be an empty set. For example, in Z^n with Euclidean metric, a sphere of radius r is nonempty only if r^2 can be written as sum of n squares of integers.

Topology

In topology, an n-sphere is defined as a space homeomorphic to the boundary of an $(n + 1)$-ball; thus, it is homeomorphic to the Euclidean n-sphere, but perhaps lacking its metric.

- A 0-sphere is a pair of points with the discrete topology.
- A 1-sphere is a circle (up to homeomorphism); thus, for example, (the image of) any knot is a 1-sphere.
- A 2-sphere is an ordinary sphere (up to homeomorphism); thus, for example, any spheroid is a 2-sphere.

The n-sphere is denoted S^n. It is an example of a compact topological manifold without boundary. A sphere need not be smooth; if it is smooth, it need not be diffeomorphic to the Euclidean sphere.

The Heine–Borel theorem implies that a Euclidean n-sphere is compact. The sphere is the inverse image of a one-point set under the continuous function $||x||$. Therefore, the sphere is closed. S^n is also bounded; therefore it is compact.

Remarkably, it is possible to turn an ordinary sphere inside out in a three-dimensional space with possible self-intersections but without creating any crease, in a process called sphere eversion.

Spherical Geometry

Great circle on a sphere.

The basic elements of Euclidean plane geometry are points and lines. On the sphere, points are defined in the usual sense. The analogue of the "line" is the geodesic, which is a great circle; the defining characteristic of the latter is that the plane containing all its points also passes through the center of the sphere. Measuring by arc length shows that the shortest path between two points lying entirely on the sphere is a segment of the great circle that includes the points.

Many theorems from classical geometry hold true for spherical geometry as well, but not all do because the sphere fails to satisfy some of classical geometry's postulates, including the parallel postulate. In spherical trigonometry, angles are defined between great circles. Thus spherical trigonometry differs from ordinary trigonometry in many respects. For example, the sum of the interior angles of a spherical triangle exceeds 180 degrees. Also, any two similar spherical triangles are congruent.

Eleven Properties of the Sphere

A normal vector to a sphere, a normal plane and its normal section. The curvature of the curve of intersection is the sectional curvature. For the sphere each normal section through a given point will be a circle of the same radius, the radius of the sphere. This means that every point on the sphere will be an umbilical point.

In their book *Geometry and the Imagination* David Hilbert and Stephan Cohn-Vossen describe eleven properties of the sphere and discuss whether these properties uniquely determine the sphere. Several properties hold for the plane, which can be thought of as a sphere with infinite radius. These properties are:

1. *The points on the sphere are all the same distance from a fixed point. Also, the ratio of the distance of its points from two fixed points is constant.*

   The first part is the usual definition of the sphere and determines it uniquely. The second part can be easily deduced and follows a similar result of Apollonius of Perga for the circle. This second part also holds for the plane.

2. *The contours and plane sections of the sphere are circles.*

   This property defines the sphere uniquely.

3. *The sphere has constant width and constant girth.*

The width of a surface is the distance between pairs of parallel tangent planes. Numerous other closed convex surfaces have constant width, for example the Meissner body. The girth of a surface is the circumference of the boundary of its orthogonal projection on to a plane. Each of these properties implies the other.

4. *All points of a sphere are umbilics.*

   At any point on a surface a normal direction is at right angles to the surface because the sphere these are the lines radiating out from the center of the sphere. The intersection of a plane that contains the normal with the surface will form a curve that is called a *normal section,* and the curvature of this curve is the *normal curvature*. For most points on most surfaces, different sections will have different curvatures; the maximum and minimum values of these are called the principal curvatures. Any closed surface will have at least four points called *umbilical points*. At an umbilic all the sectional curvatures are equal; in particular the principal curvatures are equal. Umbilical points can be thought of as the points where the surface is closely approximated by a sphere.

   For the sphere the curvatures of all normal sections are equal, so every point is an umbilic. The sphere and plane are the only surfaces with this property.

5. *The sphere does not have a surface of centers.*

   For a given normal section exists a circle of curvature that equals the sectional curvature, is tangent to the surface, and the center lines of which lie along on the normal line. For example, the two centers corresponding to the maximum and minimum sectional curvatures are called the *focal points*, and the set of all such centers forms the focal surface.

   For most surfaces the focal surface forms two sheets that are each a surface and meet at umbilical points. Several cases are special:

   - For channel surfaces one sheet forms a curve and the other sheet is a surface.
   - For cones, cylinders, tori and cyclides both sheets form curves.
   - For the sphere the center of every osculating circle is at the center of the sphere and the focal surface forms a single point. This property is unique to the sphere.

6. *All geodesics of the sphere are closed curves.*

   Geodesics are curves on a surface that give the shortest distance between two points. They are a generalization of the concept of a straight line in the plane. For the sphere the geodesics are great circles. Many other surfaces share this property.

7. *Of all the solids having a given volume, the sphere is the one with the smallest surface area; of all solids having a given surface area, the sphere is the one having the greatest volume.*

   It follows from isoperimetric inequality. These properties define the sphere uniquely and can be seen in soap bubbles: a soap bubble will enclose a fixed volume, and surface tension minimizes its surface area for that volume. A freely floating soap bubble therefore approximates a sphere (though such external forces as gravity will slightly distort the bubble's shape).

8. *The sphere has the smallest total mean curvature among all convex solids with a given surface area.*

   The mean curvature is the average of the two principal curvatures, which is constant because the two principal curvatures are constant at all points of the sphere.

9. *The sphere has constant mean curvature.*

   The sphere is the only imbedded surface that lacks boundary or singularities with constant positive mean curvature. Other such immersed surfaces as minimal surfaces have constant mean curvature.

10. *The sphere has constant positive Gaussian curvature.*

    Gaussian curvature is the product of the two principal curvatures. It is an intrinsic property that can be determined by measuring length and angles and is independent of how the surface is embedded in space. Hence, bending a surface will not alter the Gaussian curvature, and other surfaces with constant positive Gaussian curvature can be obtained by cutting a small slit in the sphere and bending it. All these other surfaces would have boundaries, and the sphere is the only surface that lacks a boundary with constant, positive Gaussian curvature. The pseudosphere is an example of a surface with constant negative Gaussian curvature.

11. *The sphere is transformed into itself by a three-parameter family of rigid motions.*

    Rotating around any axis a unit sphere at the origin will map the sphere onto itself. Any rotation about a line through the origin can be expressed as a combination of rotations around the three-coordinate axis. Therefore, a three-parameter family of rotations exists such that each rotation transforms the sphere onto itself; this family is the rotation group SO(3). The plane is the only other surface with a three-parameter family of transformations (translations along the x- and y-axes and rotations around the origin). Circular cylinders are the only surfaces with two-parameter families of rigid motions and the surfaces of revolution and helicoids are the only surfaces with a one-parameter family.

Cubes in Relation to Spheres

For every sphere there are multiple cuboids that may be inscribed within the sphere. The largest cuboid which can be inscribed within a sphere is a cube.

Permissions

All chapters in this book are published with permission under the Creative Commons Attribution Share Alike License or equivalent. Every chapter published in this book has been scrutinized by our experts. Their significance has been extensively debated. The topics covered herein carry significant information for a comprehensive understanding. They may even be implemented as practical applications or may be referred to as a beginning point for further studies.

We would like to thank the editorial team for lending their expertise to make the book truly unique. They have played a crucial role in the development of this book. Without their invaluable contributions this book wouldn't have been possible. They have made vital efforts to compile up to date information on the varied aspects of this subject to make this book a valuable addition to the collection of many professionals and students.

This book was conceptualized with the vision of imparting up-to-date and integrated information in this field. To ensure the same, a matchless editorial board was set up. Every individual on the board went through rigorous rounds of assessment to prove their worth. After which they invested a large part of their time researching and compiling the most relevant data for our readers.

The editorial board has been involved in producing this book since its inception. They have spent rigorous hours researching and exploring the diverse topics which have resulted in the successful publishing of this book. They have passed on their knowledge of decades through this book. To expedite this challenging task, the publisher supported the team at every step. A small team of assistant editors was also appointed to further simplify the editing procedure and attain best results for the readers.

Apart from the editorial board, the designing team has also invested a significant amount of their time in understanding the subject and creating the most relevant covers. They scrutinized every image to scout for the most suitable representation of the subject and create an appropriate cover for the book.

The publishing team has been an ardent support to the editorial, designing and production team. Their endless efforts to recruit the best for this project, has resulted in the accomplishment of this book. They are a veteran in the field of academics and their pool of knowledge is as vast as their experience in printing. Their expertise and guidance has proved useful at every step. Their uncompromising quality standards have made this book an exceptional effort. Their encouragement from time to time has been an inspiration for everyone.

The publisher and the editorial board hope that this book will prove to be a valuable piece of knowledge for students, practitioners and scholars across the globe.

Index

www.ingramcontent.com/pod-product-compliance
Lightning Source LLC
LaVergne TN
LVHW080727210926
848485LV00003B/529